Victorian Telegraphy Before Nationalization

Victorian Telegraphy Before Nationalization

Simone Fari
University of Granada, Spain

First published 2015 by
PALGRAVE MACMILLAN

Palgrave Macmillan in the UK is an imprint of Macmillan Publishers Limited, registered in England, company number 785998, of Houndmills, Basingstoke, Hampshire RG21 6XS.

Palgrave Macmillan in the US is a division of St Martin's Press LLC, 175 Fifth Avenue, New York, NY 10010.

Palgrave Macmillan is the global academic imprint of the above companies and has companies and representatives throughout the world.

Palgrave® and Macmillan® are registered trademarks in the United States, the United Kingdom, Europe and other countries.

ISBN: 978–1–137–40651–4

This book is printed on paper suitable for recycling and made from fully managed and sustained forest sources. Logging, pulping and manufacturing processes are expected to conform to the environmental regulations of the country of origin.

A catalogue record for this book is available from the British Library.

A catalog record for this book is available from the Library of Congress.

Contents

Acknowledgements

The research at the basis of this monograph began in June 2009, when I was studying in London on an Earth Connected Fellowship offered by the London Science Museum together with British Telecom. First of all, I would like to thank Peter Morris and the staff of the London Science Museum for having helped and encouraged me over the years. A special thanks also to David Hay and the staff of the BT library/ archives, who furnished me with precious information on telegraph company documentation. I also received helpful indications on the correspondence of the pioneers of British telegraphs during my visits to the IET archives.

As the years have passed, partial results of the research have been presented in many international conferences, thus involving a wider circle of scholars. But I would like to thank individually Graeme Gooday, Richard John, Colin Hempstead and Gabriele Balbi, who in various moments have given me much competent advice. I would also like to recall the precious exchanges with the participants at the session on telecommunications at the 2012 Economic History Society Annual Conference, at the European Fresh (Frontier Research in Economic and Social History) meeting at Pisa in the same year, the XIth Milan European Economy Workshop, the Peripheral Mobilities Workshop, Granada 2013, and the panel "Core and peripheries in the transport and communication infrastructural process" of the 41st ICOHTEC Symposium at Brasov, 2014.

Finally, I would like to thank Patricia Kennan for her painstaking care with the translation.

Introduction

> We trust that some steps will be taken either by Government or by other parties interested in the execution of such a plan, to procure the establishment of such a communication without delay, between this port and the metropolis. The advantages and conveniences of such a power of instantaneous correspondence with London would be almost incalculable. If no one else were to take the matter up, we should think it would quite repay the editors of the leading London newspapers to procure the establishment of such a Telegraph on their own account. We look forward with confidence to the time when an Electric Telegraph Office will be considered to every large town as a Post Office at the present time.[1]

Between the 1840s and 1860s the business world and press's growing demand for a swift and efficient form of long distance communication met up with a new technology offered by some innovative entrepreneurs, thus initiating the dynamic development and spread of British telegraphy. Many studies describe the origins of the service, others the players, while still others take into consideration telegraphs over the long term in Britain. This monograph presents instead an analysis of the technological and entrepreneurial features of the service, together with the companies which ran it until nationalization in 1869. It provides a historical reconstruction, mainly based on original and unedited documents belonging to a

---

[1] "Progress of Science – A New Electrical Telegraph", *Cork Examiner*, 17 May 1844.

1

variety of archives, and thus fills in numerous gaps and grey areas, so that historical facts can be interpreted with greater objectivity and accuracy. This explains the wealth of footnotes, which need to be read together with the main text.

A second feature of this monograph is that it is at the same time a study in the business and technological histories of the telegraph sector, and it aims at analysing both, as well as showing how they were continually interdependent. This hybridization is also present in the methodological approach, which keeps count of the owner- ship, the power game at the tops and financial results, and at the same time carefully tracks the technological evolution of the sector, the innovations, user needs and continuous interaction with develop- ments in other fields. Economic history, media studies and research into large technical systems converge to sustain that in network technologies like telecommunications the constitutive choices made by the principal players (entrepreneurs, engineers, politicians and users) determine a path-dependency sufficient to justify the consoli- dation of locked-in elements. In our case, each stage of evolution of the telegraph system brings with it an indissoluble tie with the choices of the past.

The first chapter illustrates the origins of British telegraphs and details the events that accompanied the Wheatstone-Cooke telegraph through the phases of invention and experiment, up to the estab- lishment of the Electric Telegraph Company. The second chapter reflects on the most salient features and dynamics to emerge from the first 15 years of activity and highlights Cooke's entrepreneurial role in the experimental phase as well as Ricardo's managerial input during the early years of the Electric. As we will show, their driving force and interaction with existing technologies went on to forge the constitutive choices from which stemmed the long-term features of the telegraph.

The third chapter focuses on the early 1850s, when the tele- graph market moved from a monopoly regime conditioned by the Wheatstone-Cooke patent to free competition. Space is given to an analysis of the strategies adopted by the new arrivals to challenge the first mover, and the latter's counter moves to foil them and hold on tight to its dominant position.

The fourth chapter covers the later part of the decade, when a series of events, partly exogenous and partly endogenous, led to the market being concentrated in the hands of two main groups. Their almost monopolistic strategies kept prices high, causing complaints and criticism to build up among their clientele. Thus they favoured in the early 1860s a new arrival, the United Kingdom Electric Telegraph Company, which proposed introducing an extremely low flat charge. Nevertheless, as described in the fifth chapter, the failure of the aggressive price war waged by the United against the dominant duopoly ended in an oligopolistic regime.

The last chapter illustrates the mechanism with which the discontent of the users, heightened by price rises following the defeat of the United, set off a political process which led in five years to the first nationalization in history, that of the British telegraph service.

1
The Origins of the Telegraph Service

10 June 1837. One of William IV's last acts was to grant a patent for Wheatstone and Cooke's telegraph apparatus, though it took ten years before the first telegraph company was authorized by Parliament, on 18 June 1846. Four years passed and it was only when serious competition began to surface that the telegraph service really began to move. Why was it so long in the making? Why was there a ten-year gap between the patent and the establishment of the first commercial company? Why did another fourteen years go by before the public began to make any significant use of it?

The few studies that have investigated the origins of the British telegraph service are generally based on the struggles to obtain patents and exclusive rights to commercial exploitation.[1] Nevertheless, in-depth research into the beginnings of British telegraphy is indispensable for defining its long-term features, which forged not only the service at home but also throughout the Empire. The political, entrepreneurial, technological and social decisions taken during the first 15 years were to function as "constitutive choices", destined to influence the mechanisms of development in British telecommunications right well past the end of the century.[2]

---

[1] John Joseph Fahie, *A History of Electric Telegraphy to the Year 1837*, London, E & F.N. Spon, 1884; Geoffrey Hubbard, *Cooke and Wheatstone and the Invention of the Electric Telegraph*, London, Routledge & Kegan, 1965; Brian Bowers, *Sir Charles Wheatstone FRS 1802–1875*, London, IEE, 2001.

[2] In this case, the term "constitutive choices" acquires a precise methodological value. It follows Paul Starr's analysis and assumes that, particularly in the field of the media, initial political and economic choices influence the long-term trajectory of a technological system. Paul Starr, *The Creation of the Media. Political Origins of Modern Communications*, New York, Basic Books, 2004.

The period of gestation can be divided into three short but clearly defined stages. The first one of "heroic" endeavour, when the telegraph apparatus was invented, went from the early 1830s to 1837, the year of the first patent. The second stage, from 1837 to 1846, was all experimentation, when the possessors of the first patent, Cooke in particular, were involved in constructing overhead lines along the railroads, with the dual aim of showing how useful they were for controlling rail traffic, and also trying them out as a potential means of communication for the public. Finally, the last "entrepreneurial" period covered the years from 1846 to 1850, with a company now authorized to transmit telegrams, though still operating on a fairly limited level.

1.1   The heroic years

Around the mid-1830s, work began on constructing telegraph prototypes, though most of the inventors involved were neither scientists nor technicians in a strict sense. Cooke, for example, was an ex-officer of the East India Company, while Morse was a fairly well-known painter,[3] Pawel Shilling a Russian diplomat[4] and Edward Davy a doctor. This was significantly reflected in the technical aspects of the early telegraphs, which were based on one or two well-consolidated scientific principles. The first needle telegraphs, like Wheatstone and Cooke's, followed on from Oersted's discovery of how electricity passing through a compass needle changes its polarity and moves it to the right or left.[5] Differently, electromagnetic telegraphs like Morse's were inspired by Arago's experiments

---

[3] Samuel Morse is commonly known as the father of the telegraph. In reality, he patented his electromagnetic telegraph in the same year in which Wheatstone and Cooke's invention was registered. Morse guaranteed for himself the exclusive right to commercial exploitation in the States, and held the license from 1846. The origins of American telegraphy are illustrated in Richard R. John, *Network Nation. Inventing American Telecommunications*, London and Cambridge, The Belknap Press of Harvard University Press, 2010, pp. 24–64.

[4] At the beginning of the 1830s, Russian Baron Pawel Shilling built a needle telegraph with applications which were initially exclusively for military use. In 1832, Shilling gave a demonstration in front of Tsar Nicholas in Berlin. When Shilling died four years later, he had managed to simplify the instrument but had not put it into practice. Most probably, a one-needle version of the latter reached Muncke, who used it in his electricity lectures. Anton A. Huurdeman, *The Worldwide History of Telecommunications*, London, J. Wiley, 2003, p. 54.

[5] Bern Dibner, *Oersted and the Discovery of Electromagnetism*, New York, Blaisdell, 1962; Ole Immanuel Franksen, *H. C. Oersted – A Man of the Two Cultures*, Birlerod,

with electricity passing through steel and magnetizing it.[6] All the telegraphs invented in this period used battery-produced dynamic electricity, an electric cell invented by the Italian Alessandro Volta at the beginning of the century. In itself the telegraph was not a sophisticated or particularly innovative contraption from a purely technical point of view. Its novelty was that it deployed scientific discoveries in electricity to transmit long distance messages at great speed. Both Morse and Cooke felt a sense of shock when, in different contexts and an almost total ignorance of the principles of electricity,[7] they realized how useful it could turn out to be for communication.[8] They were dazzled by their discoveries, and though not being members of the scientific world, poured all their intellectual, social and economic resources into realizing their inventions and exploiting them commercially.

Cooke had his moment of enlightenment in March 1836 when he witnessed a demonstration of Shilling's telegraph during a lecture given by Professor Muncke at the University of Heidelberg. The son of a famous surgeon, Cooke was in Germany to improve his technique of constructing wax anatomic models. He had noticed that his father needed models to explain human anatomy to students without having to resort to dissection, which was commonly felt to be a distasteful practice. On seeing Muncke's demonstration, however, Cooke realized that the telegraph could be turned into something of enormous social utility. He immediately threw himself

Strandsberg Forlag, 1981; Roberto de Andrade Martins, "Resistance to the Discovery of Electromagnetism: Oersted and the Symmetry of the Magnetic Field", in Fabio Bevilaqua, Enrico Giannetto (a cura di), *Volta and the History of Electricity*, Pavia/Milano, Università degli Studi di Pavia, Hoepli, 2003, pp. 245–265.

[6] Dominique François Jean Arago, "Note concernant les Phénomènes magnétiques auxquels le mouvement donne naissance", *Annales de chimie et de physique*, 32, 1826, pp. 213–223; James Lequeux, *François Arago, un savant généreux*, Paris, EDP-Sciences, 2008.

[7] In the case of Cooke, the words contained in a letter to his mother a few days after Professor Muncke's demonstration are illuminating, "I was profoundly ignorant till my attention was casually attracted to it the other day, I do not know what others may have done in the same way – this can be best learnt in London". Institution of Engineering and Technology Archive, London (henceforth IET), Sir William Fothergill Cooke Papers (henceforth WFC), SC, Mss 007, *Letter to W. F. Cooke's Mother*, 5 April 1836.

[8] This similarity between Cooke and Morse is well shown in Brian Bowers, "Early Electric Telegraphs", in Frank A.J.L. James (ed.), *Semaphore to Short Waves, Proceedings of a Conference on the Technology and Impact of Early Telecommunications* held at the Royal Society for the Encouragement of Arts, Manufactures and Commerce on Monday 29 July 1996, organized by the British Society for the History of Science and the RSA

into the task of building a telegraph and studying all its possible applications.[9] Sensing the high economic and social profitability of a new business, he abandoned wax models and went back to London at once to further his knowledge in matters of electricity,[10] work out its feasibility and assess the possible interest of backers.

Realizing he was not alone in having thought of a practical application of the telegraph, Cooke built a prototype in only three weeks with the use of a music box,[11] and turned to a watchmaker to perfection his apparatus and make it reproducible.[12] In 1836,[13] convinced he would soon have a perfectly functioning telegraph in his hands, he drew up a proper business plan,[14] in which he detailed possible applications (railways, Government, business and assistance), the low cost and efficiency as well as the technical difficulties to overcome.[15]

London, Royal Society for the Encouragement of Arts, Manufactures & Commerce, 1998, pp. 20–25.

[9] "Struck with the vast importance of an instantaneous mode of communication to the railways then extending themselves over Great Britain, as well as to Government and general purpose, and impressed with a strong conviction that so great an object might be practically attained by means of electricity. Immediately directed my attention to the adaptation of electricity to a practical system of telegraphing", William Fothergill Cooke, *The Electric Telegraph was it invented by Professor Wheatstone? Part I, Pamphlets of 1854–56*, London, W.H. Smith and Son, 1857, p. vi.

[10] "Tom and I are now going to the Adelaide Gallery to study various scientific instruments, connected more or less with our object in hand", IET, WFC, SC, Mss 007, "Letter to W. F. Cooke's Mother", 6 June 1836. Tom was William's brother.

[11] "You must know that for some weeks past I have been engaged in the construction of an instrument which I believe may prove of sufficient importance, should I succeed in bringing it to practical perfection, to merit a visit to London. Determined to satisfy myself on the working of the machinery before I went any further, I prepared to make a model, and, being unable to obtain the requisites at Heidelberg I sought them at Frankfort", IET, WFC, SC, Mss 007, "Letter to W. F. Cooke's Mother", 5 April 1836.

[12] "Whilst completing the model of my original plan, others on entirely fresh systems suggested themselves, and I have at length succeeded in combining the *utile* of each; but the mechanism requires a more delicate hand than mine to execute, or, rather, instruments which I do not possess", Ibid.

[13] "My prospectus is ready, but I am about to send it down to Mr. Chevalier, with detailed drawings, for his judgment and correction", IET, WFC, SC, Mss 007, Letter to W. F. Cooke's Mother, 21 July 1836.

[14] The term "business plan" encapsulates the form, contents and objectives of Cooke's document, though in the period it was identified with two other terms: prospectus, used by Cooke himself in the letter to his mother, and pamphlet, used by his brother in Thomas Fothergill Cooke, *Authorship of the Practical Electric Telegraph of Great Britain*, London, W.H. Smith and Son, 1867.

[15] Ibid., pp. 96–112.

However, it turned out to be more difficult to build his instrument than he had foreseen, as the clockmaker worked slowly, and the difficulties only increased as the project went ahead.[16] Cooke realized that Moore,[17] the clockmaker, knew nothing about electricity and could never have finished the job in a satisfying way.[18] At this point, armed with a business plan but held back by an unfinished instrument, Cooke was almost at the end of his tether.[19] However, he reacted quickly and set himself two objectives: look for someone with sufficient knowledge of electricity and contact possible financial backers. In November 1836 he tried to reconcile the two needs by organizing a meeting with Michael Faraday, known in the period as

---

[16] "My clockmaker has again disappointed me. I called (in Clerkenwell, nearly three miles from here) on Monday evening before coming home, in full expectation of finding that everything had been completed several days", IET, WFC, SC, Mss 007, "Letter to W. F. Cooke's Mother", 7 October 1836; "My instrument was to have been finished this morning, but upon calling I found that a wheel in the escapement movement was wrong, and had to be altered", IET, WFC, SC, Mss 007, Letter to W. F. Cooke's Mother, 22 October 1836; "Whilst the arrival of my instrument – which should make its appearance at six o'clock this evening, but which most likely will not come at all – I will employ my time in writing you. ... I can say nothing certain about my instrument at present. Moore has not behaved well, tho' he has done better the last few days", IET, WFC, SC, Mss 007, "Letter to W. F. Cooke's Mother", 17 November 1836; "My instrument came home late last night, but does not answer", Ibid.

[17] Cooke is probably referring to John Moore & Sons, the firm directed by brothers Benjamin, Richard and Josiah Moore, with a registered office in 38 Clerkenwell Close, Clerkenwell, London. According to Steve Roberts, the real maker of Cooke's first telegraph was John Brittan, an employee of the Moores'. From what emerges from the correspondence with Cooke, John Brittan carried out, always for the Moores, the first of Cooke's experiments along the London & Birmingham Railway. Steven Roberts, *Distant Writing. A History of the Telegraph Companies in Britain between 1838 and 1868*, London, Steve Roberts, 2006–2012, p. 6, unpublished in hardcopy but available online at http: //distantwriting.co.uk/, accessed 8 August 2013.

[18] "I got the best advice I could from clock and watch makers months ago, and they said they could arrange it to perfection!!! So much for their knowledge", IET, WFC, SC, Mss 007, "Letter to W. F. Cooke's mother", 17 November 1836.

[19] "My mind is nearly made up to return to modelling if the instrument does not answer.... The moment the probable fate of my instrument is decided, I will let you know. You will perceive, from what I have already said, how humbled are my (never very sanguine) expectations. I shall not give it up till all hope is gone, however. I am very anxious that it should be decided one way before I go to Treeton, as, if unfavourably, I shall make my preparation for renewing my labours in wax immediately on my return ... I am by no means disheartened at this prospect of failure, and shall set to work again, with more than former spirit, at my old occupation. I only regret having talked so much about the Telegraph. The time occupied by this affair, if it should not answer, I do not consider altogether thrown away, as I have acquired a good deal of information which will prove useful one day or other. I also have some other plans in my head, if I can find anyone to take them up, but will not venture on anything else myself unless successful here", Ibid.

"the King of Electro-Magneticians".[20] He hoped that Faraday would both give him advice over his problems and introduce him into London's scientific circles, which was a must for obtaining any financial backing.[21] All to no avail, Faraday brushed him off, saying that the scientific principles in Cooke's instrument were perfectly correct and that once completed, it could no doubt function, but that further experiments were needed to adapt it to distance communication.[22] It was tantamount to saying that Cooke's telegraph had a good potential, but Faraday himself was not at all interested in helping or backing it in any way.[23] Cooke did not lose heart, decided to change strategy and look for two figures, one to act as financial backer[24] and the other as scientific partner.[25]

He identified a possible backer, or at least a broker, in Joshua Walker, an acquaintance of his father's.[26] As a businessman and investor in railways, Walker would guarantee direct or indirect access to the world Cooke had already identified as being close to the development of telegraphy.[27] Thanks to Walker, in fact, in January

---

[20] IET, WFC, SC, Mss 007, "Letter to W. F. Cooke's Mother", 24 November 1836.

[21] Iwan Rhys Morus, "The Electric Ariel: Telegraph and Commercial Culture in Early Victorian England", *Victorian Studies*, 39(3), 1996, pp. 339–378.

[22] "He finally gave me as his opinion that the principle was perfectly correct, and seems to think the instrument capable, when well finished, of answering the intended purpose.... He said to reply to my question: 'I am afraid to inducing you by my advice to expend any large sum in experimenting, but it would be well worth working out, and a beautiful thing to carry on in this manner a conversation from distant points; and the instrument appears perfectly adapted to its intended uses'", IET, WFC, SC, Mss 007, "Letter to W. F. Cooke's Mother", 24 November 1836.

[23] Cooke wrote to his mother about Faraday's answer: "Now I consider this highly satisfactory. He took a most friendly manner, but in a way induces me to think he does not mean to take any further step in the affair. I asked his advice as to my way of proceeding in bringing it before the public when completed, but he declared his inability to advise me", IET, WFC, SC, Mss 007, "Letter to W. F. Cooke's Mother", 24 November 1836.

[24] "I do not intend to expend anything more upon it myself, but I hope to find someone who will take the risk in consideration of a fair remuneration in case of success", Ibid.

[25] "It is most probable in that case, the further prosecution of the experiments will be nearly taken out of my hands and placed in those of scientific and practical men – such, at least, I should both advise and wish", IET, WFC, SC, Mss 007, "Letter to W. F. Cooke's Father", 16 December 1836.

[26] "My object for applying to Mr. Walker being to find a person or persons connected with the busy world who could effectually bring my instrument and plans before the public, and would join me in the expense and profits", Ibid.

[27] "I thought of Mr. Walker, whose tastes, connections, situation in the neighbourhood of railroad and speculating influence...render it possible that he may interest himself about my Telegraph", Ibid.

1837 Cooke entered into contact with the directors of the Liverpool & Manchester Railway Company, as they needed a distance signal system because convoys had to be cable hauled[28] near Liverpool's Lime Street Tunnel. The company had actually already decided to adopt a pneumatic post system and stayed loyal to their first choice,[29] even though some were very much taken by Cooke's presentation. Some months later, Robert Stephenson, who had been struck by the new means, gave Cooke a second chance.[30] Together with his father, George, Stephenson was a pioneer in building engines and setting up railways companies, and was one of the most prestigious railway and civil engineers of the times. With his kind of reputation, he was clearly an influential player in the railway world.[31]

Cooke went on looking for a scientific partner, more necessary than ever since the experiments carried out in February 1837 had confirmed the problem he had detected right from the beginning (i.e., that electric signals weakened over distance).[32] Only by overcoming this limit could Cooke go on presenting the telegraph as an innovative means of distance communication. After the failure with Faraday, he turned to other scientists, until Peter Mark Roget, the secretary to the Royal Society, introduced him to Charles Wheatstone.[33] Wheatstone, who already in early youth had invented and built musical instruments, had published an essay in 1833 entitled "An account of some experiments to measure the velocity of electricity and the duration of the electric light",[34] which was so

---

[28] Steam engines were initially considered very polluting, so that to avoid them coming into the town centres, convoys were hauled in by cables, operated by a single stationary steam engine.

[29] Jeffrey Kieve, *The Electric Telegraph, A Social and Economic History*, Devon, David and Charles Newton Abbot, 1973, pp. 19–20.

[30] Morus, *The Electric Ariel*, p. 352.

[31] Michael R. Bailey (ed.), *Robert Stephenson; The Eminent Engineer*, Aldershot, Ashgate, 2003; Ken Smith, *Stephenson Power: The Story of George and Robert Stephenson*, Newcastle upon Tyne, Tyne Bridge Publishing, 2003; David Ross, *George and Robert Stephenson: A Passion for Success*, Stroud, History Press, 2010.

[32] "I tried last week an experiment on a mile of wire, but the result was not sufficiently satisfactory to admit of my acting upon it. ... Dissatisfied at the results, I this morning obtained Dr. Roget's opinion, which was favourable, but uncertain", IET, WFC, SC, Mss 007, "Letter to W. F. Cooke's Father", 27 February 1837.

[33] Ibid.

[34] Charles T. Wheatstone, "An Account of Some Experiments to Measure the Velocity of Electricity and the Duration of the Electric Light", *Philosophical Transactions of the Royal Society*, 1834, 124, pp. 583–591.

original it had won him scientific renown, followed the next year by a professorship at King's College, London. Within two years, he was elected Fellow of the Royal Society, a confirmation of his fame as a great expert in electricity.

When Cooke met Wheatstone, he was amazed. After having talked for months to scientists who were either uninterested in him or totally incompetent, he finally found himself in the company of an exceedingly well-informed expert[35] who was carrying out his own telegraph experiments.[36] Cooke realized at once that Wheatstone was the partner he had been trying to find for months,[37] while Wheatstone was in turn struck by Cooke's entrepreneurial impetus and soon convinced that he would be the ideal manager for any commercial exploitation.[38] A few days later, Cooke and Wheatstone starting working together on improving their instruments and allowing cross-contamination between their innovations.[39] Pressured by rumours in the press,[40] they deposited their *caveat* at the end of May, so that it was registered on 19 June 1837 as the first patent for an electric telegraph.[41] According to British law at the time, the patent was only provisory, and within six months the two inventors would have to deposit the specification of the instruments in question. So Cooke, and especially Wheatstone, had to work at speed to

---

[35] "The scientific men know little or nothing absolute on the subject. Wheatstone is the only man near the mark", IET, WFC, SC, Mss 007, "Letter to W. F. Cooke's Mother", 27 February 1837.

[36] "Imagine my satisfaction at hearing from him that he had four miles of wire in readiness! And imagine my dismay on hearing afterwards that he had been employed for months in the construction of a Telegraph, and had actually invented two or three, with the view of bringing them into practical use", Ibid.

[37] "Under all circumstances, I should be happy to have a scientific man for my coadjutor, though in that case I must sacrifice a larger portion of the advantages – yet I value his aid much more.... I shall be very cautious in employing the money the Doctor has so liberally placed at my command in taking out a patent, unless I see my safety very clearly, which I should do if acting in conjunction with Mr. Wheatstone", Ibid.

[38] "Mr. Cooke appeared to me to possess the zeal, ability, and perseverance, necessary to make the thing succeed as a commercial enterprise and expressed his intention of devoting his whole time and energies to it", *The Case of Professor Charles Wheatstone in the Arbitration between himself and Mr. William Fothergill Cooke*, in IET, WFC, SC, Mss 007, Box VI.

[39] Hubbard, *Cooke and Wheatstone*, p. 44.

[40] Morus, *The Electric Ariel*, p. 351.

[41] *Improvements in Living Signals and Sounding Alarums in Distant Places by means of Electric Currents transmitted through Metallic Circuits*, Patent n° 7390 (England), 10 June 1837.

make their telegraphs operate effectively by December. Whatever, they had routed the competition.

During this first, heroic phase Wheatstone and Cooke's main competitors were Edward Davy and William Alexander, a Scot. Davy was a surgeon who had been working on a telegraph prototype since 1835, totally oblivious to what other inventors were doing. By early 1837, his instrument was advanced enough to allow him to carry out practical demonstrations with a mile-long copper circuit he had been authorized to set up in Regents Park. Though his telegraph still required perfecting, in March Davy deposited a "sealed" description of his invention at the Royal Society of Arts in an effort to pre-empt Cooke and Wheatstone, amid all the rumours about their already finished and perfectly working telegraph. And in fact when Cooke and Wheatstone put in their application for a patent, Davy opposed it, but to no avail. That did not discourage him from demonstrating his instrument in several public exhibitions and working on it until he managed to deposit a patent in 1838. In the spring of the same year Davy made Cooke-like moves and contacted several financial backers close to railway and government circles. But just when a company was on the verge of being set up, the project collapsed, partly because Davy underestimated the costs and partly because of his wife's extravagance, which caused him to emigrate to Australia and leave the family to deal with the telegraph. Without the direct backing of its inventor, Davy's more complex telegraph was destined to remain unfinished.[42]

William Alexander began to promote his instrument in the summer of 1837, shortly after Wheatstone and Cooke had applied for a patent. In July 1837, he had written a letter to numerous newspapers illustrating his project for a telegraph service and adding a cost plan. A few months later he experimented with a four-mile telegraph circuit in the University of Edinburgh and later opposed Wheatstone and Cooke's request for a Scottish patent, only to withdraw because he judged his instrument inferior to theirs. In spring 1838, after solicitation, the Royal Society of Arts committee decided that his invention was nothing new. Despite the rebuff Alexander went on displaying his instrument, obviously without any hope of commercial exploitation.[43]

---

[42] On Davy's vicissitudes, see J.J. Fahie, *A History of Electric Telegraphy*; Morus, *The Electric Ariel*, 354–359.

[43] See Kieve, *The Electric Telegraph*, pp. 24–25.

In the meantime, the already-unhappy Cooke/Wheatstone collaboration took a turn for the worse, setting off a century long debate involving the two collaborators, scholars, historians and public opinion over who the inventor of the telegraph really was.[44] Those who consider the technical nature of the patented apparatus alone attribute the paternity to Wheatstone, who was known for a greater quantity of higher quality innovations.[45] Those who instead concentrate on the commercial exploitation of the invention attribute the paternity to Cooke.[46] The truth is probably somewhere between the two versions, that the contributions of the two were complementary. As Hubbard holds, Wheatstone's inventions without the entrepreneurial force of Cooke would have remained in the laboratory, while Cooke's ideas without Wheatstone's scientific skill would have been left on paper.[47]

1.2   Experiments along the railroads

10 June 1837 was a turning point in the relationship between the two inventors. The security of having a patent allowed both to dedicate themselves to the activity they were best at. Wheatstone went on in his laboratory perfecting the instrument for specification while Cooke sought possible backers and studied technical solutions for line-building. As we have noted, Cooke had already identified in railways – then in their first moment of expansion – the field of business which would allow him to raise capital quickly and make practical experiments.

Right from the beginning, the interests of the telegraph coincided with those of the Enderbys, an important family of entrepreneurs. They were London-based ship-owners who had originally made their name in whaling, but who had for many years run a factory that produced fishing and navigation goods, especially waterproof

---

[44] An extended, though somewhat biased documentation on the happenings is included in the following volumes: W. Fothergill Cooke, *The Electric Telegraph*; William Fothergill Cooke, *The Electric Telegraph: was it invented by Professor Wheatstone?*, Part II, *Arbitration Papers and Drawings*, London, W.H. Smith, 1856; T. Fothergill Cooke, *Authorship*.

[45] Bowers, *Sir Charles Wheatstone*.

[46] Tal P. Shaffner, *The Telegraph Manual: A Complete History and Description of the Semaphoric, Electric and Magnetic Telegraphs of Europe, Asia, Africa and America, Ancient and Modern*, New York, Pudney & Russell, 1859, pp. 179–215.

[47] Hubbard, *Cooke and Wheatstone*, pp. 124–125.

clothing and cables. Wheatstone, whom they had visited in his King's College laboratory, introduced Cooke to the brothers.[48] They at once expressed their interest in the telegraph, both as financial backers and suppliers,[49] and Cooke placed orders for several miles of waterproof cables, which he used to encase copper conductors.[50]

1.2.1 The London & Birmingham Railroad Company

Yet again the relationship with Joshua Walker turned out to be essential for practical experimentation. He came back into the picture, partly because of the Enderbys.[51] At the end of June, Walker met Cooke, and on the same day organized a series of meetings with the chairman and secretary[52] of the London & Birmingham Railroad Company, which needed a signal system on their new inner city railroad, where the convoys would be cable-hauled, as on the Manchester/Liverpool line.[53] In the following days Cooke explained his project to Stephenson, the company's chief engineer, already favourably impressed by Cooke's February demonstration.

In this way Cooke obtained his first official appointment: to direct the construction of the first experimental telegraph line along the

---

[48] "At King's College I was introduced to the Messrs. Enderby, who rank among the leading men in their line in London – enterprising, determined men. They were much pleased with the simple experiments shown them, and say there will be no difficulty in raising the capital, but did not offer to do it themselves – of course they must see our instruments first", IET, WFC, SC, Mss 007, "Letter to W. F. Cooke's Mother", 11 May 1837.

[49] "I heard from Mr C. Enderby yesterday that Mr. Joshua Walker seemed favourable to the undertaking, and that Mr. E. thought he would embark in it. If so we must give him good terms, as he will enter with strong interests in London and very great at Liverpool. The Enderbys are a very wealthy family, who have made most of their money in the South Sea trade. They have a large sail-cloth and rope manufactory at Greenwich, and a great many other undertakings on hand. They appear very friendly to me", IET, WFC, SC, Mss 007, "Letter to W. F. Cooke's Mother", 23 May 1837.

[50] "I am going down to Greenwich to-morrow morning to see their works, and arrange a method of covering our wires with yarn and include them in a rope, for our cross-Thames experiment. I sent the wire off yesterday. This rope, 1,500 feet in length, including 6,000 feet of wire, is to be ready by the close of the week", Ibid.

[51] Ibid.

[52] The top managerial role in the company, equivalent to today's Chief Executive Officer.

[53] "I then expressed my wish to try experiments on the railroad. 'There' he [Mr. Walker] said, 'I can at once assist you', and within half an hour introduced me to the Chairman and Secretary of the London and Birmingham Railroad. They both entered warmly into my views, and appointed the following day for a further consideration of the subject", IET, WFC, SC, Mss 007, "Letter to W. F. Cooke's Mother", 2 July 1837.

railroad between Euston and Camden. The London & Birmingham shouldered all the building costs and supplied Cooke with both the materials and the workers needed to complete the job.[54]

For the whole of July Cooke was busy building and developing the experimental line[55] and needed all his showman's skills for the numerous demonstrations he had to put on for the directors of the London & Birmingham plus their relatives and friends.[56] Initially Cooke experimented with his own telegraph, which he called an "alarum", the one he had invented in Heidelberg,[57] now with a line made with Enderby cables. The communications were elementary, based on sending electric impulses which were transformed into sounds. The demonstrations improved greatly when Cooke was able to introduce the four-needle telegraph,[58] which Wheatstone

---

[54] "To shorten details, by following up every opportunity that offered itself, and urging forward my suit unceasingly, I got through all the forms, had three interviews with Mr. Stevenson [sic.], the famed engineer, and got an order for eight hundredweight of copper wire, by Friday last; obtained leave to occupy a vast building on the railroad, 165 feet by 1000 wide, and had as many men and all the materials I could require placed at my disposal, The order was, 'Let Mr. Cooke have everything he may require'", Ibid.

[55] "I have not till now had leisure to write few lines announcing that I have completed my line of wires, extending for 13 miles, and shall have about 2½ miles more, extending to Camden Town and back, laid down as soon as I try my final experiment. I was hard at work all yesterday from a very early hour", IET, WFC, SC, Mss 007, "Letter to W. F. Cooke's Mother", 4 July 1837.

[56] "I then went home and had a good wash, dressed, with one mouthful of breakfast, and got back by 10 o'clock, the hour appointed. About 20 of the Directors were soon assembled. Mr. Wheatstone could not be present; so I commenced my explanations, and got through them with all the ease and coolness imaginable. I could not have felt less nervous had I been explaining them only to you. I prefaced by saying that I had hastened my preparations not to disappoint those Directors who were leaving town but did not offer the experiments as a sample of what my Telegraphs were to do, but simply to show that the current of fluid would pass through miles of wires instantaneously.... All expressed themselves satisfied with the principles, and seemed to take the deepest interest in the experiments", Ibid.; "Yesterday Mr. Stevenson witnessed our experiments through 19 miles of wire, extended from Euston Square to Camden Town", IET, WFC, SC, Mss 007, "Letter to W. F. Cooke's Mother", 25 July 1837; "I received a nice friendly letter from him last night, asking me to show a few experiments to two ladies staying with him, who were shortly going away", IET, WFC, SC, Mss 007, "Letter to W. F. Cooke's Mother", 29 July 1837.

[57] "I commenced by putting my Heidelberg instruments in motion, which excited great interest. I then rung a bell, &c, &c", IET, WFC, SC, Mss 007, "Letter to W. F. Cooke's Mother" 4 July 1837.

[58] Bowers, *Sir Charles Wheatstone*, p. 129. A model of this four-needle telegraph used only for experiments, can be found in the London Science Museum (LSM), *Four needles Wheatstone-Cooke Telegraph*, inventory n° 1876/1274.

was working on for the Specification. With the help of the new instrument, coded messages were sent for the first time.[59]

A hitch, though, was that the lines were made with naked copper and presented evident problems of electric dispersion, so much so that Stephenson insisted on another solution being found. Cooke worked out a new way of inserting copper conductors in tar-insulated wooden beams.[60] Insulation was increased, it is true, but so were the costs.[61]

The summer of 1837 ended in the best of ways. After the patent Cooke and Wheatstone had given public demonstrations of the functioning of their distance communication system (14 and 19 miles), to the delight of all concerned. But the autumn started with at least two ugly surprises for Cooke. The first was that the press attributed the invention and installation of the telegraph almost exclusively to Wheatstone,[62] and while Cooke had spent the whole of July installing

---

[59] "He [Mr. Stephenson] wishes us also to have all our instruments on the most approved construction, and I have consequently put several new ones in hand, to be ready, if possible, in a fortnight. He said we must have two or three rehearsals beforehand, that all may go off in 'good style'. He declared himself a 'convert to our system', and seemed quite delighted at the correspondence we carried on at so great a distance from each other, requesting me to send the word 'Bravo' along the line more than once. It ended by his desiring me to send an invitation to Mr. Wheatstone to join us, which he politely replied to by saying he would do himself 'the honour'", IET, WFC, SC, Mss 007, "Letter to W. F. Cooke's Mother", 25 July 1837.

[60] "He [Mr. Stephenson], begged me to lay down my wires permanently between those two points on my best plan, with a view to extending the communication hereafter, if the Directors approved.... He was allowed us to go to another £100 expense on account of the Company, and only added as a condition that all should be perfect, for 'I do not like to be laughed at for a failure'. ... I have just given orders for 5,000 feet of wood to be sawn in a particular manner, with grooves for the wires, which I am going to have boiled in coal tar previously to laying down. Our wire is all ready", Ibid.; LSM, *Original telegraph cable to Euston Square and Camden*, inventory n° 1867/37.

[61] "He [Mr. Stephenson] was prepared to recommend a much more expensive method of protecting the wires, which would ensure more thoroughly their safety, the cost of which might come to £400 or £500 per mile, or £100.000 to Liverpool; that, though the sum was great, yet the Telegraph was of vital importance to the railroad", IET, WFC, SC, Mss 007, "Letter to W. F. Cooke's Sister", 8 September 1837.

[62] In July 1837, the news of the invention and public demonstrations of the telegraph was published in the British newspapers, but most of them did not give the names of the two inventors, and where they did they gave only that of Wheatstone. "New and beautiful invention", *Morning Post*, 1 July 1837; "New and beautiful invention", *Reading Mercury*, 1 July 1837; "Electric Telegraph", *Fife Herald*, 6 July 1837; "Extraordinary and beautiful invention", *Preston Chronicle*, 15 July 1837; "Electric Telegraph", *Gloucester Journal*, 22 July 1837.

and implementing the Euston-Camden line,[63] his name was not even mentioned in the papers. A compromise was quickly found. Cooke would act as sole manager of the telegraph, dealing with its installation and commercialization in Great Britain, for which he would receive 55% of the revenue from the patent. The other 45% was left to Wheatstone, who would pocket the installation rights from the continent.[64] In this way, the existing situation was given a formal blessing. Cooke could tranquilly dedicate himself to his role as entrepreneur and Wheatstone to working on the Specification.

The second ugly surprise was that in spite of their initial enthusiasm, the London & Birmingham directors preferred to install a pneumatic signal system.[65] After the positive results obtained in both the July and September experiments, Cooke seemed convinced the company would adopt his telegraph, even more after the changes in wire-laying advised by Stephenson.[66] However, the sharp rise in costs

---

[63] "I saw a most incorrect account of our proceedings in the paper the other day, in which Mr. Whetstone and Mr. Stephenson are represented as trying the experiments, which have been conducted solely under my direction, I alone being in communication with the Directors on the subject, and have been the only person responsible from the commencement. I have good reasons for taking no notice of the newspaper reports for the moment. Do not fear that I shall have my full credit at last", IET, WFC, SC, Mss 007, "Letter to W. F. Cooke's Sister", 8 September 1837.

[64] "I went down by appointment to Mortlake last night with Mr. Wheatstone, and discussed with Mr. Hawes [the arbitrator] the heads of our agreement. We were occupied at it from 8 till 11 o'clock, when, everything being settled to our mutual satisfaction, our signatures were attached, and the papers placed in lawyer's hands to embody their spirit in legal circumlocution. They contain all I wish, leaving me sole and entire manager in England, Scotland and Ireland, with this one exception – that before selling the patents or licenses, I am to obtain Mr. Wheatstone's acquiescence in the price. Nothing, of course, can be more reasonable. I am to have a percentage of 1–10th for my expenses and trouble, dividing the balance equally with Mr. Wheatstone, so that out of every £100 I shall have £55 and Mr. W. £45. ... I leave to Mr. W the right of introducing the invention into Belgium, Holland, Prussia and France, for his exclusive benefit, I having the same privilege in Russia and Austria", IET, WFC, SC, Mss 007, "Letter to W. F. Cooke's Sister", 9 November 1837. These arrangements are described in greater detail in Articles 4 and 6 in *Heads of Agreements between W. F. Cooke and Charles Wheatstone*, November 1837, IET, WFC, SC, Mss 007, IV, 7.

[65] "Electricity was thought of as a quicker signal agent, and some successful experiments were tried with it, but experience has proved that the whistle is more advantageous and suitable at every respect", *Osborne's Guide to the London and Birmingham Railway*, 1840, quoted in Kieve, *The Electric Telegraph*, p. 28.

[66] "Mr. Stephenson not only approves, but will recommend Cooke's and Wheatstone's Telegraph to the Directors for adoption. ... Mr. S. comes down to-morrow evening, when he is to show me his report, and has offered to make any alteration I may suggest in the plan he recommends to the Directors. All, therefore, is going on as well as can be desired", IET, WFC, SC, Mss 007, "Letter to W. F. Cooke's Sister", 8 September 1837.

after the modifications probably convinced the directors to adopt a well-tested and cheaper signal system. There just remained the problem of the already-installed line, which the railway company had financed and legally owned. Cooke was reluctant to lose the fruit of his labour and agreed to buy the line at a half of the installation cost.[67] This enabled him to go on using it both to improve his installation techniques and display the running operations of his system to possible new purchasers.

1.2.2 The Great Western Railway Company

A third business opportunity came to Cooke from the railways. Early in 1838,[68] Isambard Kingdom Brunel, chief engineer of the Great Western Railway Company, offered Cooke the chance to experiment with his telegraph along the new railway line between London and Bristol.[69] Brunel was a many-sided genius, a civil and mechanical engineer who developed numerous projects including a tunnel under the Thames, the Great Western railway line, the first ocean steam liners, and a prototype of a pneumatic railway. Like Stephenson, he followed family tradition and in the 1830s already had a great reputation as one of the most brilliant men of the times.[70]

Unlike the first two railways the Great Western had no specific need for distance communication because it used single engines on double tracks. Most probably it was Brunel's idea to install a telegraph line because, like Stephenson, he was fascinated by this new technology and wanted to take part in its development. It was, in fact, Brunel who asked Cooke to insert the insulated copper conductors in steel tubes

---

[67] "The L. and B. [London and Birmingham] Railway Company have offered me the instrument (used at the experiments) at half price, and declared, when twelve in council, the most friendly feeling both towards me and the undertaking", IET, WFC, SC, Mss 007, "Letter to W. F. Cooke's Mother", 20 December 1837.

[68] In reality, the first informal contacts had already taken place in late 1837: "I have arrived at nothing definite with Brunel, but a prospect seems to be opening soon his line (the Great Western), IET, WFC, SC, Mss 007, "Letter to W. F. Cooke's Mother", 2 December 1837. According to Roberts, it was Stephenson who presented Brunel to Cooke. Roberts, *Distant Writing*, p. 7.

[69] Brunel, who had already realized the importance of the telegraph, was also courting Edward Davy. But the latter patented his instrument almost a year later than Wheatstone and Cooke, without adding any significant innovations for users (it was still a six-wire needle model). For technical and legal reasons, Brunel preferred to draw up an agreement with Cooke. Morus, *The Electric Ariel*, pp. 355–359.

[70] Adrian Vaughan, *Brunel – An Engineering Biography*, Surrey, Ian Allan Publishing, 2006; Annabel Gillings, *Brunel. Life & Times*, London, Haus Publishing, 2006.

to protect them from the weather and any acts of hooliganism.[71] The line between Paddington and Drayton was set up between June 1838 and June 1839, financed by the Great Western, which employed Cooke as construction manager with a retribution of £165 a mile. Between 1839 and 1842, it was used almost exclusively for the Great Western.[72] Then, in 1842, the company announced it was calling a halt to the project and not continuing the line as far as Bristol. Cooke made a drastic decision and offered to finance the line up to Slough and repair the damaged part between Paddington and West Drayton. Slough was very near Windsor Castle, while Paddington was within easy reach of Whitehall and the government offices, which fitted in nicely with Cooke's original idea of the telegraph serving the Government. It was indeed a clear investment in the future.[73] Furthermore, as Cooke was using his own money, he made a wise reduction in fixed costs by replacing his four-needle telegraph,[74] which required six wires, with the two-needle model, which worked on only three, and substituting Brunel's costly underground steel tubes with naked overhead wires simply suspended from wooden poles. He obtained a patent for this system, which acted as a model for many others, thanks to its enormous reduction in building costs in comparison to underground lines.[75] As proprietor of the line,

---

[71] "Our plans are entirely changed; and I am to lay down the wires in iron pipes", IET, WFC, SC, Mss 007, "Letter to W. F. Cooke's Mother", 8 June 1838.

[72] "Electric Telegraph. – This extraordinary machine is now being worked on the Great Western Railroad, between Drayton and Paddington; and though no distinct idea of the apparatus can be imparted without plans and draughts of the dial, pipes, rods, &c. of which it is composed, yet the principle will excite unqualified admiration when our readers learn that intelligence is conveyed at the rate of two hundred thousand miles per second, or eight thousand times quicker than light travels during the same period, by means of electrical currents passing through coils of copper wire, placed immediately behind some fine magnetic needles, made to operate upon a circular series of twenty letters, which indicate such terms, either separately or collectively, as they have been arranged to represent. This telegraph will act both day and night, in all states of the weather, and with a rapidity so superior to the common process, that one minute only is required for the communication of thirty signals", "Electric Telegraph", *Leicester Chronicle*, 29 August 1840.

[73] Hubbard, *Cooke and Wheatstone*, p. 102.

[74] The telegraph used by the Great Western had four needles and not five as some authors affirm. Confirmation in William Fothergill Cooke, *Telegraphic Railways; or the Single Way, recommended by Safety, Economy, and Efficiency, under the Safeguard and Control of the Electric Telegraph*, London, Simkin, Marshall & C., 1842, pp. 14–15.

[75] *Improvements in Apparatus for transmitting electricity between distant places*, Patent n° 9465 (England), 8 September 1842.

Cooke licensed the running of the service to Thomas Home, who thus became the first British telegraph operator. The Paddington office opened its doors to the public in 1842, for all to witness transmissions or send telegrams for the unit cost of a shilling.[76]

1.2.3   The London & Blackwall Railway Company

Meanwhile, in autumn 1839, shortly after work had finished on the Paddington-West Drayton line, Cooke received an interesting offer from another firm, the London & Blackwall Railway Company,[77] which had commissioned Stephenson, together with George Bidder, to construct a three-and-a-half mile cable haulage system between the Minories (near the Tower of London) and Blackwall, the busy steam ship port further down the Thames. As the system was activated by a stationary steam engine positioned at the terminus, it needed a signal system to notify the operator of the presence of passengers on the trains. Unlike the Liverpool & Manchester and London & Birmingham, which both used cable haulage systems, the distance in this case (six stations) was too long, so Stephenson recommended the electric telegraph immediately.[78] Starting from the five-needle telegraph patented in 1837, Cooke then created a simple and inexpensive system of railway signalling. Each of the terminus stations was equipped with a five-needle telegraph with autonomous needles connected to a single needle installed in each intermediary station, which moved left to indicate "stop" and right to indicate "go on".[79] The system took only a few months to install, and in the spring of 1840 the Minories-Blackwall line was inaugurated, with Wheatstone-Cooke's telegraph as an essential device for traffic control and passenger safety.[80]

---

[76] Kieve, *The Electric Telegraph*, pp. 32–33.

[77] "I have greater pleasure in writing to you now, as I can announce the improved prospects of the Telegraph, Mr. Robert Stephenson having engaged me to lay it down on the London and Blackwall Railway", IET, WFC, SC, Mss 007, "Letter to W. F. Cooke's Sister", 28 November 1839.

[78] Among the enthusiastic backers of the telegraph was also Peto, one of the three contractors for the London-Blackwall. Roberts, *Distant Writing*, p. 7.

[79] Cooke, *Telegraphic Railways*, pp. 12–15.

[80] "London and Blackwall Railway", *Morning Chronicle*, 6 July 1840. "We must not omit to mention briefly the beautiful apparatus to Messrs. Wheatstone and Cook (sic.), by which instantaneous communication is effected between the terminal stations, or between any one station and any other on the line. Without this, one of the most splendid inventions of modern times, the working of the Blackwall Railway according to the present system would have been rendered rather hazardous", Francis Whishaw, *The Railways of Great Britain and Ireland*, London, John Weale, 1842, p. 269.

Great prominence was given in the press to the reports of the Select Committee on the railways, which had been set up to regulate the booming service.[81] The fifth report in particular gave ample space to the telegraph as necessary for guaranteeing punctuality and safety. Though it gave due space to the telegraph's government and business functions,[82] it dwelt mostly on its role in flanking the railroads.[83] A later pamphlet by Cooke entirely dedicated to telegraph systems along railroads pointed out not only the possibility of guaranteeing punctuality and safety, but also of drastically reducing construction and maintenance costs.[84]

Following on from Cooke's observations and the experiences of the Great Western and the London-Blackwall, Bidder, an engineer for the latter and also a member of Parliament, defended the telegraph-supported single-track system during the parliamentary debate over authorizing the construction of the Yarmouth-Norwich line.[85]

For the first time, the telegraph was described as an indispensable control/safety system for the railways drastically cutting fixed costs, thanks to the move from two tracks to one. It was also perceived to

---

[81] "Electric communication", *Reading Mercury*, 11 July 1840; "Taxation of Railways – Electrical Telegraphs", *Northampton Mercury*, 18 July 1840; "Electrical Telegraph", *Bath Chronicle and Weekly Gazette*, 23 July 1840; "Electrical Telegraph", *Royal Cornwall Gazette*, 31 July 1840; "Electrical Telegraphs", *Chelmsford Chronicle*, 21 August 1840; "Fifth report of the Select Committee on Railways", *Yorkshire Gazette*, 22 August 1840; "Fifth report of the Select Committee on Railways", *Preston Chronicle*, 22 August 1840.

[82] "If the new telegraph of which I [Wheatstone to the Select Committee on the Railways] have spoken succeeds – and it has succeeded perfectly so far as experiments have been tried – we might place three telegraphs in connexion with the six wires now used on the Great Western Railway; and these might be applied as I have said before, to three specific purposes – one exclusively for railway purposes; another, to be let to any persons who choose to avail themselves of it; & another for Government objects", "The Electrical Telegraph", *Bath Chronicle and Weekly Gazette*, 20 August 1840.

[83] "There is one suggestion with respect to the use of the telegraph for railroad purposes which should not be overlooked, being of the greatest importance, inasmuch as all danger from collision would be obviated, and more prompt assistance rendered in case of accident. Mr. Wheatstone's proposal is to have posts through which the magnetic wires can be carried up, and with an apparatus on the top placed at every quarter of a mile along the line. By this means the guard having with him a portable instrument, might communicate a message in either direction of the line at pleasure", Ibid.

[84] Cooke, *Telegraphic Railways*.

[85] "I consider it to be fair and reasonable; and I will undertake to find parties, undeniably competent, to take the railway off our hands at the sums I have stated to the Committee. The estimate is made for a single line; the single lines are found to answer with proper regulations. The electrical telegraph is intended to be applied to his line, and is included in the estimates; no accident can occur except from wilful neglect; the traffic is of the simplest character", "Yarmouth and Norwich Railway Bill", *Norfolk Chronicle*, 23 April 1842.

cut service costs, since a punctual timetable would lead to a rationalization of both personnel and fuel. In August 1842, thanks to Bidder's
support, Cooke presented the advantages of the telegraph to the
shareholders of the Yarmouth-Norwich Railway Company, showing
how the telegraph was no longer confined to the offices of engineers
or company managers, but was becoming of public dominion.[86]
Two years later, in May 1844, the Norwich-Yarmouth railway started
operating, with the electric telegraph as one of its most innovatory
features. It was emerging that the telegraph not only made railway
transport cheaper, safer and more reliable, but it also offered new
opportunities for the business-minded.[87]

However, the Select Committee and the ensuing fame of the telegraph helped worsen the already-strained relations between Cooke
and Wheatstone. The latter had been the central figure in the committee's hearings and had been received and listened to as though he
were the only inventor involved. The newspapers publishing the
committee's reports also made much mention of the professor from
King's College and mostly passed over his partner in silence. Cooke's
irritation grew, increasing with Wheatstone's constant reminders

---

[86] "The chairman wished to call upon another gentleman to give his opinion as to the
working of the line which was about to be worked on almost, but not altogether a new
principle. Mr. Cooke was the patentee of the electrical telegraph, and would be able
to tell them how beautifully it could be worked", "Shareholders in the Yarmouth and
Norwich Railway Company", *Norfolk Chronicle*, 13 August 1842.

[87] "The adoption of the Telegraph on this Line has effected a material reduction in
the engine and carriage department, and will certainly be a source of income to some
considerable extent, as an instantaneous mode of transmitting shopping and other
commercial intelligence. As a further practical benefit conferred by the Telegraph unre-
mitting vigilance and alertness may be enforced, upon all the servants of the Line, by
the instant and infallible detection at the terminus of individual remissness. While travelling will be safe, because a train cannot arrive without notice: the men must be on the
alert, because a message may be received and require an answer at any moment. Another
great advantage arising from the application of the Telegraph to this Line, consists in
the more perfect adaptation of the Railway to the wants of the country through which
it passes. On market-days passengers may be collected in horse carriages, dispatched
for them from the nearest stations; and agricultural produce may be carried in wagons
along the Railway, in the intervals between the trains. All such minor sources of traffic
well deserve the attention of this Railway, in proportion as its cheapness of construction will enable smaller returns to realise a larger proportional dividend. Numerous
little rills and streamlets will swell the tide of traffic, and the Road-side population will
for the first time participate in the convenience of the vast works which have deprived
them of their local conveyances. In short this Line will become, what all were once
intended to be, highway roads open to the use of the public", "The Electric Telegraph
on the Yarmouth & Norwich Railway", *Norfolk Chronicle*, 4 May 1844.

that their telegraph was basically his with a few small improvements, with the result that the two of them ended back with the arbitrators.[88] Wheatstone named as his arbiter John Frederick Daniell, Professor of Chemistry at King's and an authority on electro-chemistry, while the highly respected Marc Isambard Brunel, the father of Isambard Kingdom, was Cooke's choice. The final award went again in favour of the status quo, with Cooke to be considered the practical promoter of the telegraph, crucial for its development and commercialization, while Wheatstone was to be recognized as its inventor.[89] Apparently, the result of the arbitrage contented both men on the surface, though their personal diatribe went on for many more decades.[90]

1.2.4 The Electric Telegraph Company

By the beginning of 1843 Cooke had obtained excellent results with the railroads and his independent undertaking with the Great Western. He now received authorization to construct what he had long planned, a government line between Slough and Windsor. He decided therefore to accelerate procedures for fully exploiting the telegraph patents. The first step was to remove Wheatstone as part owner. Knowing his partner's natural aversion to any kind of risk, Cooke let it be known he was in financial difficulty because of enormous infrastructural expenses incurred for the Great Western.[91] As a result, he felt there were only two options: involve Wheatstone

---

[88] For a description of the events leading to the arbitrage, see Cooke, *The Electric Telegraph*.

[89] "While Mr. Cooke is entitled to stand alone, as gentleman to whom this country is indebted for having practically introduced and carried out the Electric Telegraph as a useful undertaking, promising to be a work of national importance; and Professor Wheatstone is acknowledge as the scientific man, whose profound and successful researches had already prepared the public to receive it as a project capable of practical application; it is to the united labours of two gentlemen so well qualified for mutual assistance, that we must attribute the rapid progress which this important invention has made during the five years since they have been associated", "Award of Sir Marc Isambard Brunel and Professor Daniell", in T. Cooke, *The Authorship*, p. XXIX.

[90] The arbiters' decision had little coverage in the press, though it can be found in "Messrs. Cooke and Wheatstone's Electric Telegraph", *Bury and Norwich Post*, 19 May 1841.

[91] "At the beginning of the year 1843 we were at our lowest point of depression. The patents remained almost unproductive, and we had incurred, in various ways, a considerable outlay", Cooke, *The Electric Telegraph*, p. 213. This comment inserted by Cooke in his pamphlet against Wheatstone is said to have been taken, in his own words, from a letter he sent to Wheatstone in early 1843. Some authors interpret his comment literally, holding that in 1843 Cooke had strained resources and was near bankruptcy.

financially or have Wheatstone sell his share of the patents, so that Cooke could sell them on to backers.[92] Initially reluctant, Wheatstone gave way and agreed to sell his part,[93] and on 12 April 1843 he ceded his rights on the patents jointly owned with Cooke[94] in exchange for royalties on every line built, according to the distance covered.[95]

Cooke's timing was perfect because telegraphs grew in importance in 1844, and in August he obtained the commission for a 160-mile line, the longest in existence up to then. The Admiralty had decided to substitute the optical telegraph between London and Gosport (alongside the naval port of Portsmouth) with a modern electric system. Cooke was to construct two linked branches (London-Southampton and Southampton-Gosport) along the railroads of the London & South Western Railways Company. It was the most generous contract Cooke had ever signed, with the Government paying £24,000 for the construction of the line and an annual

---

(See Roland Wenzlhuemer, *Connecting the Nineteenth-Century World. The Telegraph and Globalization*, Cambridge, UK, Cambridge University Press, 2013, p. 170). However, partly because of how events turned out, it can be reasonably surmised that Cooke was bluffing to obtain Wheatstone's share of the patent rights. On a later occasion, Cooke reused the statement to emphasize his own entrepreneurial role in the first stages of the telegraph.

[92] "Having come to the conclusion that unless, with our own capital, we put up specimen telegraph on the suspension plan, our disappointments might be endless, I had already obtained from the Great Western Railway Company the privilege of putting up and working for a term of years a telegraph to Windsor. It had been proposed that Mr. Wheatstone should either find half the capital required for this Windsor telegraph, or cede to me as a 'separate privilege' the right of putting it up on my own account. This I was willing to do in aid of the common enterprise, though I had no capital conveniently at command, and could not hope for a remunerative return", Cooke, *The Electric Telegraph*, p. 214.

[93] "He [Wheatstone], declines (I do not say improperly, for it appeared that we had misunderstood each other on a collateral point) both my alternative propositions. What was to be done? To stand still was ruinous, and Mr. Wheatstone was not willing to move in the direction which seemed likely to open better prospects. Numerous meetings took place between us at the office of Mr. Wheatstone's solicitor, who happily for my credit, was not then unwell....In January 1843, it was definitely proposed at an interview at Mr. Richardson's [Wheatstone's solicitor] office, that Mr. Wheatstone should assign to me his share in the patents for a fixed royalty", Ibid., pp. 214–215.

[94] The text of the agreement is in IET, WFC, SC, Mss 007, VI, 14.

[95] "Mr. Wheatstone's royalty would be a charge upon the joint Property. Its amount is as follows. For the first 10 miles of Telegraph laid down every year £20 per mile. For the second 10 miles £19. For the third 10 miles £18. For the fourth £17. For the fifth 10 miles £16. For all miles above 50 miles £15. Mr. Wheatstone has also a licence for the use of the inventions for certain domestic and other purposes, limited to distances of half a mile as the extreme extent of his communication", IET, WFC, SC, Mss 007, II, 121, *Statement respecting the Electric Telegraph*, 9 May 1843.

£1,500 for maintenance and management. Partly for this reason Cooke decided for the first time to appoint a construction manager and chose Owen Rowland, who finished the job in early 1845.[96] The press reacted for the first time with enthusiasm,[97] listing the military[98] and commercial benefits of the new means and foreseeing a great future for it nationwide.[99]

Initially little interested in the telegraph, the dailies now added the hype, geared on by stories that caught the public's interest, like the birth of Queen Victoria's second child[100] or the capture of a murderer.[101]

Thus, a virtuous circle was established, with the press promoting existing lines, which created new requests by the railway companies. They, in turn, attracted by the potential costs reduction, sought to introduce the new signalling instrument. In September 1845, for

---

[96] "South Western Railway", *London Standard*, 24 August 1844; "London and South Western Railway General Meeting", *Reading Mercury*, 31 August 1844; "Electric Telegraph on the South Western", *Chelmsford Chronicle*, 31 January 1845; "Electric Telegraph on the South Western", *Reading Mercury*, 1 February 1845.

[97] For a good example of technological enthusiasm, see "The century of inventions", *Glasgow Herald*, 10 February 1845. There were some sceptics, though a clear minority: "All the world's a Bedlam. An old Gentleman's opinion of things in general", *Leicestershire Mercury*, 8 March 1845.

[98] "There is little doubt that they will be extended to all the more vulnerable extremities of the land; and it is easy to see how vastly this beautiful invention, combined with railways, will add to the security of the Kingdom, both from foreign invasion and domestic insurrection. The electric wires, extending over the island, may be compared to the nerves ramified over the body, which gave instant notice of the slightest movement in the most distant member. The government seated in the sensorium will enjoy, when danger threatens, a sort of omnipresence. It will be able to communicate with the remotest parts in a few seconds, to know what is passing in these parts and to direct, without the loss of an instant, the measures which the conjuncture requires. The danger known, the railway furnish immediate and gigantic powers to meet it. With their aid a march, which in former times occupied a month, is contracted to a day; and supposing ten thousand soldiers to be stationed in London, they could now be sent to York in less time than would have been spent on the march to Windsor seven years ago", "The Electric Telegraph", *Manchester Courier and Lancashire General Advertiser*, 15 February 1845.

[99] "In cases where express accounts of public events are required, the invention promises to be invaluable, since it is calculated by the patentee that were a telegraph to be laid down between Liverpool and London, accurate reports of speeches might be transmitted from either place, and be put in type and circulated long before any express train could possibly make its arrival", "Electric Telegraph on the South Western", *Hampshire Telegraph*, 1 February 1845.

[100] "Birth of a Prince", *Leeds Intelligencer*, 10 August 1844.

[101] "The Cords that Hung Tawell", in Kieve, *The Electric Telegraph*, pp. 29–46.

example, the South Eastern Railway drew up a contract with Cooke for the telegraph on all its railroads, for a total of 124 miles. Given the amount of work, Cooke entrusted the setting up of the lines to an outside contractor, W.T. Henley. Work on the South Eastern network was finished in July 1846, coinciding with the official establishment of the Electric.[102]

The growth of telegraph lines between spring 1843 and late 1845 was awesome. When Wheatstone exchanged his rights for royalties, there were only 60 miles of the telegraph lines in Great Britain along the railroads of three companies. In November 1844 there were over 250 miles of telegraph lines,[103] and by 1845, 500 miles.[104] Such a development obviously meant a proportional increase in Wheatstone's royalties, which were becoming more and more burdensome for Cooke.[105] To save the situation, he thought to facilitate the entry of new partners, and in July 1845 began working on Wheatstone to get him to cede his residual rights for a one-off payment. The negotiations over Wheatstone's golden handshake went on until the December of the same year,[106] when he agreed to a pay-off of

---

[102] Roberts, *Distant Writing*, p. 13.

[103] "The Electric Telegraph has been adopted on the following lines:
  – By order of the Lords of the Admiralty on the South Eastern Railways as a Government Telegraph from the Admiralty Whitehall to Portsmouth above **90 miles**;
  – On the same line as a Commercial Telegraph from Nine Elms to the port of Southampton **77 miles** – with a branch to Gosport **15 miles** on the London and Blackwall Railway;
  – Great Western Railway from London to Slough **18 miles** – The Windsor Telegraph;
  – Yarmouth and Norwich – a "single line" of Railway – **20 miles**;
  – Part of the Oldham Branch Railway;
  – Part of the Leeds and Manchester Railway;
  – Part of the Edinburgh and Glasgow Railway;
  – The Dalkey & Atmospheric branch of the Dublin and Kingston Railway".
  IET, WFC, SC, Mss 007, II, 125, *Draft advertisement for the Electric Telegraph, 2 November 1844*, (underlining in original, bold added by the author).

[104] "THE ELECTRIC TELEGRAPH is already established, or in course of formation, to an extent exceeding 500 miles, upon the following Metropolitan and other Lines of Railway. METROPOLITAN LINES: SOUTH EASTERN, SOUTH WESTERN, GREAT WESTERN, LONDON AND BIRMINGHAM, LONDON AND CROYDON, LONDON AND BLACKWALL (about to be united by a Branch with the Eastern Counties Railway). COUNTRY LINES: SOUTH DEVON, YARMOUTH AND NORWICH, MAIDSTONE AND TUNBRIDGE, NORTHAMPTON AND PETERBORO, NORTH MIDLAND, SHEFFIELD AND MANCHESTER, PRESTON AND WYRE, LEEDS AND MANCHESTER, EDINBURGH AND GLASGOW", IET, WFC, SC, Mss 007, II, 133 (small caps and bold present in original).

[105] Wheatstone pocketed £444 for the 1844 royalties, but as much as £2,775 for 1845, almost seven times as much. Data in IET, WFC, SC, Mss 007, II, respectively 122 and 152.

[106] The letters on this negotiation are in IET, WFC, SC, Mss 007, VI, 1–11.

£30,000.[107] Though the sum was decided by Wheatstone himself,[108] it did not stop him from regretting it within a few months.[109]

While busy negotiating with Wheatstone, Cooke also took steps to achieve two other objects. On 2 September 1845 he provisionally registered a joint-stock enterprise named the Electric Telegraph Company,[110] to manage patents and administer already installed lines.[111] At the same time he conducted a long negotiation with two possible financial partners, Bidder and John Lewis Ricardo.[112] The former had already played a crucial role for Cooke in the construction of telegraph lines along the London & Blackwall and Yarmouth & Norwich railways, while the latter was a high flyer in London business circles, a member of Parliament since 1841, and a scion of the historic family of brokers. His grandfather, Abraham, had not only founded the family firm but had become the chairman of the Stock Exchange.[113] After numerous informal and formal meetings,[114] on 28 November 1845, Cooke, Bidder and Ricardo signed a preliminary agreement for the sale of some hefty quotas of telegraph patents.[115] Beyond the subdivision of the rights to the patents, the agreement provided for the partnership to be named the "Electric Telegraph Company" and for the signees to commit themselves to sustaining the expenses for a formal recognition by Parliament, the condition that would allow the entry of new partners and the start of commercial activity.[116] With the signing of the appropriate contracts, the November agreements were formally recognized on 23 December 1845:[117] Bidder and Ricardo bought 23/32 of the rights

---

[107] Ibid.

[108] See particularly IET, WFC, SC, Mss 007, II, "Letter from C. Wheatstone to W.F. Cooke", 2 August 1845.

[109] Cooke, *The Electric Telegraph*, pp. 89–90.

[110] IET, WFC, SC, Mss 007, II, 134, *Certificate of Registration of the Electric Company*.

[111] IET, WFC, SC, Mss 007, II, 135–139.

[112] IET, WFC, SC, Mss 007, II, 140 e 143–151.

[113] David, the famous economist, had also worked in the family firm, but had married a Quaker and given up the religion of his family, who for this reason broke off with him. Piero Sraffa, Maurice Dobb (editors), *The Works and Correspondence of David Ricardo, Biographical Miscellany*, Vol. 10, Indianapolis, Liberty Fund, 2005.

[114] IET, WFC, SC, Mss 007, IV, 8–9.

[115] IET, WFC, SC, Mss 007, IV, 10, *Articles of Agreement respecting the sale of Letters patents and Co-partnership and Company to work the patents*.

[116] "The style of the said co-partnership shall be 'The Electric Telegraph Company'", IET, WFC, SC, Mss 007, IV, 10, *Articles of Agreement respecting the sale of Letters patents and Co-partnership and Company to work the patents*, Articles 14 and 21.

[117] IET, WFC, SC, Mss 007, IV, 10, *Articles of Agreement respecting the sale of Letters patents and Co-partnership and Company to work the patents*, Articles 1–10.

on the telegraph patents, for the total sum of £115,000, while Cooke kept 9/32 of the ownership, which all came to a total of £160,000.[118] Initially Wheatstone was appointed scientific consultant to deal with research submarine cables. But the dispute he had with Alexander Bain, an inventor who accused him of having appropriated a patent for an electric clock, force the Electric to acquiesce to the House of Lords' ruling, indemnify Bain with Wheatstone's shares and force Wheatstone out of the company.[119]

With Bidder and Ricardo as co-proprietors of the telegraph patents, a new phase in British telegraphs began. After being the prime player in the heroic and experimental phases, Cooke cut himself out a secondary role as a major shareholder and director, enjoying the economic advantages of having sold the rights but also giving up the enormous responsibility of being sole manager. Bidder and Ricardo, both members of Parliament and railway men, represented the direct and indirect interests of three parties, which up to then had been out on the margins of the telecommunications sector – politics, big railway business and finance.

1.3   The dynamics of the Electric

After being presented in Parliament in February, the Bill for the recognition of the Electric became an Act of Parliament and was officially signed by Queen Victoria on 18 June 1846.[120] The absence of any official debate both inside and outside Parliament[121] suggests that there was no question of it not being approved, and probably steps had been taken to protect it.[122] The Bill was, for example,

---

[118] IET, WFC, SC, Mss 007, IV, 11–12.

[119] In reality, the incident has never been fully clarified. Wheatstone said he had never formally accepted the post of scientific consultant to the Electric, and consequently had never received any salary from them. Cooke, on the contrary, always maintained that Wheatstone had been taken on with the precise task of experimenting submarine cables, but was then dismissed because he was held responsible for the court case brought by Bain.

[120] *An Act for forming and regulating "The Electric Telegraph Company", and to enable the said Company to work certain Letters Patents* [18 June 1846], 8 Victoria, c. XLIV.

[121] The absence of a parliamentary debate is noted in "The Electric Telegraph Company Bill", *London Standard*, 17 February 1846; "House of Commons", *London Standard*, 3 March 1846.

[122] See the defence of the Bill in Parliament moved by Hawes, who since July 1846 had been a shareholder of the Electric. Hawes held that the Electric Telegraph Company Bill was a strictly private provision, and for that reason deserved a quicker procedure. But some, like Roebuck, felt it was a public act because the company wanted to draw up contracts with the railways. HC *Deb*, 6 April 1846.

presented as a parliamentary authorization for a new company and, as was usual in the period, its internal organization was regulated in great detail[123] – rights,[124] duties,[125] and in general, relations with institutions[126] and creditors.[127] For this reason, the Electric Telegraph Company Act had the format of a standard document, very much like those presented for the recognition of various railway companies. Only a few articles were in fact dedicated specifically to regulating the service and foresaw: free access on equal terms;[128] the possibility of building and running an exclusive service for Her Majesty's Government;[129] the possibility for the Government to seize temporarily all telegraph lines for reasons of public order and national security.[130] Unlike the railways, the telegraph service would be regulated exclusively by laws which would from time to time authorize new companies as well as others which more generally regulated the conduct of any public service.[131]

Following parliamentary approval of the company with unlimited liability,[132] the first shareholders' meeting of the Electric was held on 1 July 1846. Of the 6,000 shares available, 4,480 were subscribed, for the total sum of £448,000. The main single shareholder was Bidder with 1,540 shares, followed by Cooke, who took the sizeable quota of 1,160 shares.[133] Though Ricardo held only 728 individual shares, his

---

[123] Articles IV, XII–XV, XIX–XXV, *An Act for forming and regulating "The Electric Telegraph Company", and to enable the said Company to work certain Letters Patents* [18 June 1846], 8 Victoria, c. XLIV.

[124] Ibid., Articles XXVI–XXXIV, LIII–LV.

[125] Ibid., Articles IX–XI, LI–LII.

[126] Ibid., Articles XXXV–XLI, LVI.

[127] Ibid., Articles XXXV–XLI, LVI.

[128] Ibid., Article XLIV.

[129] Ibid., Article XLV.

[130] Ibid., Article L.

[131] In more detail, the telegraph service was regulated by *An Act to attach certain Conditions to the Construction of future Railways authorized or to be authorized by any Act of the present or succeeding Sessions of Parliament; and for other Purposes in relation to Railways* [9 August 1844], 7–8 Victoria, c. LXXXV; *An Act for consolidating in One Act certain Provisions usually inserted in Acts authorizing the taking of Lands for Undertakings of a public Nature* [8 May 1845], 8 Victoria, c. XVIII; *An Act for consolidating in One Act certain Provisions usually inserted in Acts with respect to the Constitution of Companies incorporated for carrying on Undertakings of a public Nature* [8 May 1845], 8 Victoria, c. XVI.

[132] In those times, British legislation was still reluctant to acknowledge limited liability for shareholders. Although it was foreseen by the law, it was a long and difficult process to obtain it, so that a company with a capital structure like the Electric was unlikely to ask for it. For greater detail, see Paul Johnson, *Making the Market. Victorian Origins of Corporate Capitalism*, Cambridge, Cambridge University Press, 2010, pp. 137–166.

[133] British Telecom Archives, London (henceforth BT), *Electric Telegraph Company*, TGA 1/6, *Shareholder books 1846–1856*. 1 July 1846.

family group held 1,456.[134] With two minority holdings,[135] Thomas Boulton, a city broker, and Benjamin Hawes, member of Parliament and brother-in-law to Brunel, completed the picture of the eight shareholders. In spite of this impressive line up of new faces, the board of directors reflected the key players in the preliminary agreement – Ricardo, chairman, with Cooke and Bidder, and Samson Ricardo as auditor.[136] This leads us in part to reconsider the positions of some authors who hold that the Electric was founded by a syndicate of railway engineers who had bought the rights from Cooke.[137] In reality, it was Cooke who sought and found other financial partners in Bidder and Ricardo, with whom he shared the ownership of the patents and transferred it to the Electric. So the first board of the Electric witnessed continuity with the trio of the agreement, and the limited shareholder structure shows the absence of other strong members of the railway world, with the exception of Bidder and Ricardo.

From the managerial, economic and core business points of view, the first five years of life of the Electric can be divided into two distinct periods: 1846 and 1847, which featured a substantial continuity with Cooke's individual management style, and 1848–1850, which marked a clean cut with the past and a new management model.

1.3.1 Ties with the past

Cooke initially was left in charge of the construction of the lines and basically went on doing what he had been doing since 1837,[138] making agreements with railway companies to build telegraphs lines alongside the railroads. This activity, which had increased

---

[134] The Ricardo shareholdings were distributed as follows: John Lewis 728, Samson, uncle and partner, 616, Albert 56 and Frederick 56.

[135] Benjamin Hawes held 100 shares, Thomas Boulton 224.

[136] BT, *Electric Telegraph Company*, TGA 1/3/1, *Board Meeting Book*.

[137] "Along with some friends and a large group of railway engineers Ricardo was part of a syndicate that purchased the electric telegraph patents from Cooke and Wheatstone in October 1845. ... The consortium consisted of a large group of railway engineers focused on Robert Stephenson, and a smaller group focused on Ricardo", Roger Neil Barton, "New Media. The Birth of Telegraphic News in Britain 1847–68", *Media History*, 16/4, 2010, pp. 381–382; "On September 3, 1845 a syndicate led by the Ricardo family of City merchants projected a joint-stock company to purchase all the patents Cooke and Wheatstone had obtained to date and provide capital for their more effective working, particularly to gain an income from public messages through a national network of telegraph lines. This created the *Electric Telegraph Company*", Roberts, *Distant writing*, p. 14.

[138] Ibid.

considerably since 1844, was, in quantitative terms, the core business of the Electric, as is shown by the fact that during this period, the company's main income came from building and running telegraph lines for third parties.[139] Constructing private lines, however, meant that at the same time the Electric was able to develop its own business network, since wires for its exclusive use were always laid alongside those for the railways. Private telegraph correspondence started operating officially on 1 January 1848,[140] and until then did not seem to be of much importance either for the public or the Electric. The railway companies themselves undervalued the importance of private telegram messages, and started drawing up conventions over the division of the revenue only two years later.[141]

By the end of 1846 the Electric possessed an extended but basically unorganized network. With 1,048 miles of lines, it was still formed of independent unlinked circuits, the most important of which were the 180-mile-long Eastern Counties Railway and the 250-mile Midlands Railways.[142] However, signs of a change in strategy surfaced in 1847, announced in late October 1846 by press reports of the Electric's intention to build a new central station in London, which would connect up with all London stations and consequently all the cities in the network.[143] Cooke had already rented some rooms to use as offices and stores for Electric in the Strand, a prime city location, and an ideal choice for a central station.[144] Evidently priorities changed, for the Electric finally positioned its station facing the Bank

---

[139] Though not particularly analytic, the first known figures for Electric (December 1847) chalked up a credit of £338,270. BT, *Post Office*, POST 30/202C, *Statements of Accounts 1847*.

[140] "On Saturday [1st January] the principal central telegraph station of the Electric Telegraph Company in Moorgate Street, opposite the Bank of England, was publicly opened for the transmissions of messages to all parts of England", "Opening of the Central Telegraph Station in the City", *London Daily News*, 3 January 1848.

[141] "York & North Midland Railway. Conclusion of the third report of the Committee of Inquiry", *York Herald*, 3 November 1849; "Eastern Counties Railway Company", *Chelmsford Chronicle*, 23 April 1850.

[142] Henry Tuck, *The Railway Shareholder Manual, or the Practical Guide to all the Railways in the world, completed, in progress, and projected; forming an entire Railways Synopsis indispensable to all interested in railway locomotion*, London, Effingham Wilson, 1848, p. 155, p. 163.

[143] "Subterranean Electric Telegraph through the Metropolis", *London Standard*, 20 October 1846.

[144] "During the last few weeks considerable interest has been excited in the scientific world and the several railway companies whose lines run into the metropolis, by the announcement that the Electric Telegraph Company intend forthwith to establish a central telegraph station, at the company's depot in the Strand, by means of which communication will be obtained from one point to all parts of the country", Ibid.

of England, in a more convenient area for telegraph cables coming from the railway stations, which were themselves the terminals of numerous intercity circuits.[145] The location was considered more important than the building itself, given that the Electric initially seemed to opt for rooms in the Royal Exchange, then the seat of Lloyds,[146] but later bought Founders' Hall, a completely renovated building in nearby Lothbury Street, a part of the space is still known today as Founders Court.[147]

After the opening of a central station in London, others followed in quick succession – Manchester in March,[148] Liverpool[149] and

---

[145] "The Secretary said that the company intended to bring the electric telegraph into the City by the way of London-bridge, King William-street, and Princess-street, into Lothbury, where the terminus of the company would be, and from thence they intend to proceed along Moorgate-street into Finsbury. The object of the company was to bring the electric line of the South Western and Dover Railways by the above line into the City; and from the terminus at Lothbury to unite them with the Eastern Counties, the London and Birmingham, and the Great Western lines, so as to form one general communication with the different railways throughout the Kingdom", "The Electric Telegraph Termini", *Morning Post*, 17 March 1847.

[146] "Intended central termini of the Electric telegraph at the Royal Exchange", *London Standard*, 19 November 1846.

[147] "On Monday, the whole of the extensive buildings, including Founder's Hall and Chapel, in Founders-court, Lothbury, fronting the Bank of England, were being demolished, the Electric Telegraph Company having purchased the property for the formation of their central metropolitan station", "Telegraphic Central Station", *Chelmsford Chronicle*, 14 May 1847.

[148] "We mentioned a few weeks since the intention of the London Electric Telegraph Company to establish an office, in connection with the various lines of their telegraph on the different railways, in some central part of Manchester, and their desire that it should be within the Exchange premises, if practicable. We now learn that the agent of the Telegraph Company for this part of the Kingdom, waited on the Exchange committee, and applied for a room for the purpose of the telegraph, which the company were desirous to rent from the committee, if a suitable apartment could be had, and the terms could be arranged. The Exchange committee showed Mr. Cox plans of the extension of the Exchange-street, 100 feet in length and 30 feet wide. The committee named the rent they should require for such room, when erected, and also intimated that in the meantime they could place at the disposal of Mr. Cox the present committee-room, – and Mr. Cox said he would lay the proposal before the Telegraph Company. In all probability, we may expect to have the central telegraph station, to which the electric wires will converge from all parts of the kingdom, in the new Exchange building", "The Electric Telegraph for Manchester", *Blackburn Standard*, 24 March 1847.

[149] "A gentleman was in Liverpool, on Saturday last making arrangements on the part of the Electric Telegraph Company for introducing that wonderful mode of communication between this town and the metropolis. He had previously made similar arrangements in Manchester. As the same company has already communications with the most other parts of the Kingdom, Liverpool will be at once connected with a great variety of places, including most of the sea ports on the south and western coast", "Introduction of the Electric Telegraph into Liverpool", *London Standard*, 14 April 1847.

Leeds in April,[150] and Newcastle in May.[151] Wherever in the provinces the Electric opened a central station, the local business elite showed interest and came to its aid. In Manchester it obtained rooms in the new seat of the Exchange under construction.[152] In Liverpool[153] and Newcastle[154] offices were found in the local exchanges, in Leeds in the Commercial Building, where "the commercial community ... w[ould] then be placed in as good a position to derive advantage from one of the most wonderful inventions of modern times".[155]

With the creation of offices no longer dependent on the railways, the Electric carried on building lines, and by mid-1847 there were 1,200 miles operating and another 1,000 to be shortly completed and/or installed.[156] Evident not only the increase in mileage, but also the strategic importance of the new lines. Direct connections between London, Birmingham, Manchester and Liverpool,[157] as well as the London/Edinburgh/Glasgow link-up,[158] aimed at attracting a mainly commercial and financial clientele. In August 1847 the Electric unveiled its intention to start a message service that would be open to the public, which would begin in the New Year.[159] By means of

---

[150] "We are informed that the Electric Company have taken offices in the Commercial Buildings, to which it is their intention to extend their telegraphic wires", "The Electric Telegraph in Leeds", *Leeds Mercury*, 24 April 1847.

[151] "The Electric Telegraph Company have taken offices in the Central Exchange, at Newcastle-upon-Tyne, which will be a great accommodation to the inhabitants of that town", "The Electric Telegraph", *Leamington Spa Courier*, 15 May 1847.

[152] This passage from the official discourse on the laying of the first stone of the new Exchange is significant: "Actuated in these views, he was happy to tell them that one facility, and a very important one for the transaction of commercial business arising from that important invention the electric telegraph, was now about to be connected with this Exchange. (Hear, hear) The Electric Telegraph Company were now putting down their wires, so that long before the building was complete the subscriber would be able to come here and hold communication not only with other merchants of his own neighbourhood, but the whole of England!", "Enlargement of the Manchester Exchange", *Manchester Times*, 21 May 1847.

[153] "The Electric Telegraph in Liverpool", *Freeman's Journal*, 20 November 1847.

[154] "The Electric Telegraph", *Leamington Spa Courier*, 15 May 1847

[155] "The Electric Telegraph", *Bradford Observer*, 15 July 1847.

[156] "Electric Telegraph Company", *Stamford Mercury*, 30 July 1847.

[157] "The posts are all laid down along the London and North Western line, and the wires are in course of adjustment. In the course of a few weeks the whole will be completed, and then there will be perfect telegraphic communication between London, Birmingham, Manchester, and Liverpool", "The Electric Telegraph", *Worcestershire Chronicle*, 22 September 1847.

[158] "The Commercial Electric Telegraph Company", *Stirling Observer*, 7 October 1847.

[159] "Electric Telegraph Company", *Manchester Times*, 28 August 1847.

an intensive campaign of press announcements, the Electric let it be known it was opening subscription rooms in its central offices with a collecting service for economic and financial news.[160] As the newspapers observed, "the Electric...intend establishing the 'Commercial Telegraph' for the exclusive use of mercantile men in London and the chief towns in the kingdom, distinct from the telegraph reserved for the use of railways and the Government".[161]

Nevertheless, intentions are not always the same as results, and although the Electric was investing energy and money, the commercial service only came into being at the end of 1847. There was no organization of personnel to back the service, the central offices were often unlinked from the railway stations and it fell to porters to guarantee the flow of the telegrams.[162] Furthermore, although private correspondence was allowed in theory, in the first 18 months of the Electric's life there was so little that it did not appear as a single entry source of income in the half-yearly balance sheet.[163] In conclusion, at the end of the first two years' activity, the Electric survived economically thanks to its work building lines for third parties but, aware that the card up its sleeve would be developing telegraph correspondence and circulating intelligence, it was dedicating a large part of its economic and human resources to it.

1.3.2 The venture

In 1848 with the start of the correspondence and intelligence distribution services, the organizational structure of the Electric was modified with its new core business in mind. A decentralized organization was established in six geographic areas corresponding to the six most important circuits,[164] but hierarchically under the chairman, Ricardo, who regulated both the correspondence and intelligence services with daily orders for the management and staff, on the model used by the

---

[160] Ibid.

[161] "Electric Telegraph Company", *Leicester Journal*, 24 September 1847.

[162] "The arrangements for communicating with distant places, by telegraph, are at present so very different from the plan about to be adopted, that we cannot judge of the effect the one by the other. Now a person can only communicate with opposite parts of the country by visiting the different telegraph stations at the several railway termini in London", Ibid.

[163] BT, *Post Office*, POST 30/202C, *Statements of Accounts 1847*.

[164] In the first half of 1848, the six districts were (1) Central London, (2) Manchester, (3) Eastern Counties, (4) Midland, (5) Scottish, and (6) York, Ibid.

early railway companies.[165] Immediately under the chairman was the secretary, William Henry Hatcher, who also covered the post of chief engineer. At the head of each district there was a superintendent, who had to answer for his budget to the chairman and board of directors. The aim was to improve the productivity of all districts by investing the local managers with responsibility. The circulation of news service was put under the control of the Intelligence Department, which was entrusted to James Acland, a journalist evidently chosen because he had worked with Ricardo on the Anti-Corn Law Campaign. However he soon had to be dismissed because of his involvement in an illegal sale of intelligence, and his place was taken by Charles Boys, who held the post until the nationalization of the telegraphs. Initially, the mostly financial intelligence was sent to the subscription rooms and was only available to paid-up members, usually from organizations like stock exchanges, Royal Exchanges and Mechanics Institutes. According to Barton, it was probably Ricardo himself who introduced this service, which he used regularly as a member of the London Stock Exchange. Yet it was not a great success, partly because subscribers often exploited it by selling or circulating the news acquired to the newspapers. Unable to put a stop to this piracy, the Electric intro-duced a direct sales service to newspapers and institutes and ended by closing the subscriptions room (1849), though the Intelligence Department operated right through to nationalization.[166]

Despite all the investments and organizational adjustments carried out in the first half of 1848, the results of the new service were far from encouraging, as all the districts recorded net oper-ating losses,[167] so much that in April 150 members of the staff were dismissed.[168] The troubles were not only economic,[169] for complaints and criticism began to arrive from the new clients, especially the

---

[165] BT, *Electric Telegraph Company*, TGA 1/2, *Chairman's order book 1848–1850*.

[166] Barton, *New Media*, pp. 383–386.

[167] BT, *Post Office*, POST 30/202C, *Statements of Accounts 1848*.

[168] "The London Electric Telegraph Company, we are sorry to see, does not find it so advantageous a speculation as was at first expected; and last week they gave notice to 150 employees, mechanists, &c., that their services would be dispensed with for the present. Should the Company not be better supported, in all probability a great number more will be discharged, the expenses being so considerable, nearly 3000 persons being employed on the different lines and the London offices in various duties", "The Electric Telegraph", *Dundee, Perth and Cupar Advertiser*, 14 April 1848.

[169] E.C. Baker, *Sir William Preece, FRS, Victorian Engineer Extraordinary*, London, Hutchinson & Co., 1976, pp. 49–50.

press, who judged the cost of telegrams too high, both in an absolute sense and in comparison to similar services in the United
States and the German-speaking countries.[170] As some journalists
pointed out,[171] the higher tariffs in Britain were partly due to the
initial fixed costs, the widespread use of underground lines in built-
up areas, which were far more expensive than the overhead system

---

[170] "The price charges appeared to give considerable dissatisfaction", "Opening
of the Central Telegraph Station in the City", *London Daily News*, 3 January 1848;
"Every encouragement should be given to telegraph companies to lay down wires, for
although we have got to a certain stage of advancement, the electric telegraph system
in this country is far from being in a satisfactory state. It seems very desirable that it
should not be left a monopoly in the hands of the Electric Telegraph Company, or the
Government, who, by inveterate adherence to one system, may check the course of
improvement. The use of the needle telegraph by the company we believe to be fraught
with great inconvenience, and, indeed, in particular conditions of the weather, as the
needle telegraph will not work, it may become useless either to announce an invasion
or to communicate orders. It is to be observed that the electric telegraphs for the south
coast are in a bad condition. The coast line is not completed, and the South Devon
line is said to work imperfectly. The telegraph on the South Eastern is worked in a
complicated manner; there is no telegraph on the Brighton. There is a telegraph on the
South-Western but on the Great Western none beyond Slough. We say nothing about
military communications with the inland stations, or with Chatham, Plymouth, and
Milford. All this requires looking to, so that every encouragement be given to complete
the system; and in case of need, the government must themselves lay down wires",
"Our national defences", *Sussex Advertiser*, 8 February; "We deeply regret to be obliged
to stigmatise the manner in which the Electric Telegraph Company's officers carry
on the business connected with this wonderful and highly important invention. To
advertise 'instantaneous communication' and yet take 6½ hours to transmit a message
of sixty words is really too bad. It is high time that something should be done to curtail
the efforts of these avaricious monopolists, and that the enormous charges which they
make should be brought to a reasonable and fair standard. As matters at present exist,
the public are outrageously taxed; their business, in many instances, impeded; and
the utility and value of the Electric Telegraph itself, in a great utility and value of the
Electric Telegraph itself, in a great degree, unappreciated", "The Americas", *Glasgow
Herald*, 9 April 1849; "Electric Telegraphing in England advertised to be instantaneous,
accurate, and cheap", *London Standard*, 28 April 1849; "Not only, however, has the
public to complain of manifold delays and inaccuracies, but also of the exorbitant rates
charged for the transmission of messages. 'Carpe Diem' appears to be the motto of the
company, and whilst there is no competition, the disposition to act up to it is fully
developed. Such a state of affairs cannot, however, be suffered in a commercial community", *Newcastle Journal*, 2 June 1849.
[171] "The cost per mile for the construction of the English telegraph has hitherto varied
from 120£ to 150£ (the line originally laid down on the Great Western Railway being
300£ per mile). In America, on the other hand (where, indeed, the system is extremely
crude and imperfect), the cost is only from 20£ to 30£ per mile; and in Prussia the wires
are conducted under-ground for 40£ per mile, and on the American plan for 20£. This
disparity, as might have been expected, affects very materially the respective rates of
charge for the transmission of messages", "The Electic [sic] Telegraph", *Dundee, Perth,
and Cupar Advertiser*, 21 December 1849.

used outside the big cities.[172] In fact, because of the high level of urbanization, Parliament had obliged the Electric to lay the cables underground, in order to facilitate the circulation of people and goods.[173] The problem evidently did not exist outside towns, where the Electric was able to build along the railroads, which were private property anyway, but in the high-density urban areas it was forced to go underground to reach the central offices. This was the strategic reason why in 1848 only 10 of the 266 offices were located outside railway stations.[174] Savings, yes, but in the second half of 1848, it was clear that the fixed costs had soared well over the initial previsions and that a hefty injection of capital was essential.[175]

The entry of new shareholders was crucial, and in December 1848 the number rose from 12 to 27, more than doubling in little more than a year, and with them came new capital for the company. However, more important than the actual number of new entries was that among them there were high flyers from the political and business worlds, especially the railways, who would inevitably modify the ownership of the Electric.

George Wilson had been the first important shareholder to join the original shareholders in December 1847 with a minority holding

---

[172] Consider, for example, just two of the numerous descriptions of cabling in 1847 and 1848. On the central London station: "On Saturday the conducting pipes of the Electric Telegraph Company, opening the communication between the central station in Lothbury, now rapidly approaching completion, and all places south of the metropolis, were being laid down in King William street and Princes-street. This company will open direct lines of communication from this station to fifty of the principal towns in England during the ensuing three months", "Electric Telegraph" *Morning Post*, 6 September 1847. On the Manchester seat: "The wires will be conveyed below the pavement of the streets in a two inch pipe, and brought up into the room from the basement story of the building", "The Electric Telegraph for Manchester", *Blackburn Standard*, 24 March 1847.

[173] "And be enacted, that it shall be lawful for the Company from Time to Time to lay down and place under any Street, and either along or across such Street, but not above the same, any Pipes or Tubes, not being of larger Size than Three Inches Bore, which shall or may be necessary or convenient for the Purposes of any Electric Telegraph or intended Electric Telegraph, or for conveying the conducting and other Wires of such Telegraph along or across such Street", Article XXXV, *An Act for forming and regulating "The Electric Telegraph Company", and to enable the said Company to work certain Letters Patents* [18 June 1846], 8 Victoria, c. XLIV.

[174] Barton, *New Media*, p. 382.

[175] The chairman himself of the *Electric*, Ricardo, in order to justify the high tariffs, listed the heavy costs for infrastructures and personnel. "The Electric Telegraph Company", *The Times*, 27 April 1849.

of 100 shares.[176] He was an industrialist from Manchester, known for his role in the Anti-Corn Law League and therefore politically close to Ricardo, but when he joined the Electric, he was also director and deputy chairman of the Manchester & Leeds Railway.[177] Nevertheless, the first big shake-up came in December 1848, when five important railwaymen came into the company: Thomas Brassey (333 shares), Samuel Morton Peto (125), Richard Till (100), David Waddington (70) and Joseph Paxton (50).[178] Thomas Brassey was considered the top railway contractor and builder in the period; by 1848, he had built about a third of the railways in Britain and before his death in 1870, 20% of railways elsewhere in the world.[179] For his railways in Britain, he had sought the collaboration of two of the most famous railwaymen of the period – Brunel and Stephenson, both advocates of the telegraph right from its earliest stages. Brassey also worked with other contractors, especially Peto, who happened to become an Electric shareholder in the same period. Before turning to railway contracting, Peto had been a very successful builder and by the time he joined the Electric, he was also a member of Parliament.[180] Richard Till, Peto's lawyer and partner in numerous railway contracts, was another shareholder. He also held appointments as director of the Norfolk Railway and secretary of the London and Birmingham Railway.[181] David Waddington, an iron and steel industrialist from Manchester and a member of Parliament since 1847, held an important managerial post as vice-chairman of the Eastern Counties Railways when he became an Electric shareholder.[182] The last of this first wave of new shareholders was Joseph Paxton, the famous landscaper, who was later commissioned for the

---

[176] BT, *Electric Telegraph Company*, TGA 1/6, *Shareholder books 1846–1856*, 1 July 1847.

[177] Charles William Sutton, "Wilson, George (1808–1870")", in *Dictionary of National Biography, 1885–1900*, London, Smith, Elder and Co., Vol. 62.

[178] BT, *Electric Telegraph Company*, TGA 1/6, *Shareholder books 1846–1856*, 1 December 1848.

[179] Arthur Helps, *Life and Labours of Mr. Brassey*, London, Bell and Daldy, 1872; Tom Stacey, *Thomas Brassey: The Greatest Railway Builder in the World*, London, Stacey International, 2005.

[180] Edward C. Brooks, *Sir Samuel Morton Peto Bt. 1809–1889. Victorian Entrepreneur of East Anglia*, Bury, Bury Clerical Society, 1996; John G. Cox, *The Achievements and Failings of a Great Railway Developer*, London, Railway & Canal Historical Society, 2008.

[181] Robert Colley, "Railways and the Mid-Victorian Income Tax", *Journal of Transport History*, 24/1, 2003, pp. 78–102.

[182] G. Dalling, "David Waddington. A Great Survivor", *The Great Eastern Railway Society Journal*, 19/20, 1978, pp. 20–21.

Crystal Palace project,[183] but in 1848 he was still a director and share-holder of the Midland Railways.[184]

Two other top railway men were registered during the following shareholders' meeting – Stephenson (175 shares), who was by then a member of Parliament, and John Hawkshaw.[185] The latter was chief engineer of the Lancashire & Yorkshire Railway, and was destined to become one of the most highly regarded British civil engineers.[186] Later, in December 1849, Henry Arthur Hunt was registered. An engineering inspector, he had had building contracts with many railway companies and in 1847 had designed Founders' Hall.[187]

During 1849 another shareholder joined the group, though this time not strictly linked to the railway world. William Wylde, with his 50 shares, was for different reasons considered worthy of note.[188] He was an artillery colonel who in 1847 had been appointed Groom of the Bedroom by Prince Albert. And for this reason some authors think he was the court's man inside the Electric, where he became a director, regardless of his minority share quota.[189]

The new ownership structure brought about changes at the top. In March 1849, Richard Till, Peto's right-hand man, appeared on the board of directors.[190] Some months later, on 8 November, Cooke resigned from his directorship, at the same time as Hawes, who had however never taken part in the meetings.[191] Their exit almost coincided with the ingress of Wylde and Hunt.[192] The misunderstandings with Cooke, of an unknown nature, were resolved and he came back onto the board in January 1850, this time with Peto at his side,[193] though probably without the great influence he had wielded until late

[183] Kate Colquhoun, *A Thing in Disguise. The Visionary Life of Joseph Paxton*, London, Harper Perennial, 2004.
[184] Asa Briggs, *Victorian People*, London, Pelican Books, 1970, pp. 43–44.
[185] BT, *Electric Telegraph Company*, TGA 1/6, *Shareholder books 1846–1856*, 30 June 1849.
[186] Hugh Chisholm, "Sir John Hawkshaw", *Encyclopaedia Britannica*, Cambridge, Cambridge University Press, 1911.
[187] Hunt had also worked as an inspector with the North Staffordshire Railway, London, Brighton and South Coast, Eastern Counties, District and Metropolitan. "Sir Henry Arthur Hunt (1810–1889), Obituary", Institution of Civil Engineers, Minutes of Proceedings, 1890.
[188] Ibid., 31 December 1849.
[189] Roberts, *Distant Writing*, p. 44.
[190] BT, *Electric Telegraph Company*, TGA 1/3/1, *Board Meeting Book*, 2 March 1849.
[191] Ibid., 8 November 1849.
[192] Ibid., 21 November 1849.
[193] Ibid., 10 January 1850.

1848. In the same period Hatcher, who had also been present since the establishment of the Electric, was dismissed, first of all as secretary (replaced by Fourdrinier) and later as chief engineer (replaced by Edwin Clark, one of Stephenson's men). Though the reasons for Hatcher's dismissal and Cooke's exit from the board of directors are unknown, they are most likely to have been due to the new shareholders,[194] who may well have wanted to make the Electric more dynamic and therefore opted for more proactive, more go-ahead experts less identified with the early telegraphs.

The new line-up of shareholders registered a clear majority of businessmen or technicians with railway interests. The influence of railwaymen had appeared right back with the preliminary agreements, where two of the three founding shareholders, Bidder and Ricardo, were railway managers. Yet during 1849 there was a change in direction. Right in the moment of economic difficulties, the most influential men in the railway world committed themselves personally to bringing in capital and know-how. Unfortunately, the reasons for their actions do not emerge clearly from known sources, and therefore only surmises can be offered. Firstly, the most important railway companies had telegraph communication systems maintained and extended by the Electric. Secondly, most of the railwaymen were heavyweight businessmen who needed a distance communication service, the monopoly of which would be held by the Electric for another two years. Thirdly, with the prospect of new competition, which had been talked about for some time, the strengthening of the ties between railway companies and the Electric would guarantee, at least to the former, advantageous conditions for the maintenance and management of private networks as well as exclusivity in building telegraph lines alongside the railroads.

The new-entry shareholders brought about a decisive change in financial results. In the two halves of 1849 the Electric chalked up profits from its commercial activity and was able to distribute dividends.[195] The positive trend continued the next year[196] and brought the board to ask for a modification to the 1846 Electric Telegraph

---

[194] Hatcher's resignation is not recorded in the board's minutes, but it can be surmised, given that the directors refer both to a new secretary and a new chief engineer. Ibid., Minutes 1849–1850.

[195] BT, *Post Office*, POST 30/202C, *Statements of Accounts June 1849* and *December 1849*.

[196] Ibid., *Statements of Accounts June 1850* and *December 1850*.

Company Act.[197] To increase the number of shareholders, the directors proposed modifying the capital stock, transforming a single £100 share into more attractive £25 ones.[198] The starting point of the operation was that up to 1850 only half of the capital had been paid up and therefore only £50 for every £100 share.

Four £25 shares were given to existing shareholders for all shares with a £100 nominal value. Of the four shares, £20 needed to be paid up for the first two, and only £5 for the other two. What it meant was that a shareholder who could not pay off the whole sum could quite freely get rid of the two shares with the higher pay-off, and so keep the cheaper two.[199]

The overall aim of the operation was to rationalize the capital and increase its value. The new Act, which actually foresaw the possibility of increasing capital immediately with shares with a pay up of £2 and 10s, was approved in the first half of 1851,[200] and resulted in a prodigious increase in the number of shareholders, 69 strong in December 1851.[201] While the 1849 entry had brought in powerful shareholders, the 1851 one opened the door to many small investors.[202] With the split-up of its capital, the Electric had received the consecration of the market and could at that point concentrate its efforts on expanding the service and developing strategies for defeating its rivals, who in the meantime had entered the market once the telegraph patent's 14 years had expired.

---

[197] *An Act for amending the Act relating to the Electric Telegraph Company* [24 July 1851], 14–15 Victoria, c. LXXXVI.

[198] "The Electric Telegraph Company", *Liverpool Mercury*, 22 February 1850.

[199] BT, *Electric Telegraph Company*, TGA 2/1, *Half Yearly Report and Balance Sheet 1851–1870*, June 1851.

[200] Ibid., December 1851.

[201] BT, *Electric Telegraph Company*, TGA 1/6, *Shareholder books 1846–1856*, 31 December 1851.

[202] In 1850–1851, some other important figures came in, like Edward Riley Langworthy (100 shares in December 1850), who was a Liberal politician in the cotton business, George Edward Dering, who was an inventor in the field of the telegraph and electricity (50 shares in July 1851) and Edwin Clark himself, who became chief engineer of the Electric. From December 1852, the splitting up of the capital was a consequence of the entry of ordinary investors who bought small quantities (5,10, 20) of the new issue. Ibid.

2
Constitutive Choices

The driving force, the main player in all three initial phases of British telegraphs, was Cooke, who right from the beginning operated as an entrepreneur in the Shumpeterian sense of the word. He identified an innovative technology, studied its possible applications to industry and commerce, raised the necessary capital and then founded the first company in the sector, laying the basis for a long dominion with the help of the entry barriers that he, too, had raised. Only at the end of this operation, its survival and industrial continuity guaranteed, did Ricardo step in. His great merit was to identify in correspondence and intelligence services the core businesses of the Electric and forge its organizational structure to meet the need.

Through the three evolutionary stages described in the previous chapter, the telegraph took on its own distinctive features, thanks especially to the entrepreneurial force of Cooke and Ricardo, and to the way it both imitated and innovated the technologies employed by other services of the times. The electric telegraph emerged as the low-cost heir to the optic version, living in symbiosis with the railways, absorbing their capital and know-how and at the same time copying gas pipe techniques. It was held back however by the rival Penny Post's universal circulation and the need to win the competition with the pneumatic post for long-distance communication. From this interaction between technologies old and new, the long-term features of British telegraphs emerged.[1]

---

[1] Gabriele Balbi and Cecilia Winterhalter (eds), *Antiche novità. Una guida transdisciplinare per interpretare il vecchio e il nuovo*, Napoli-Salerno, Orthotes Editrice, 2013.

2.1 "Ne tentes aut perfice"[2]

What helped to make Cooke the most salient figure in the first ten years of British telegraphs was the tenacity with which he appropriated an innovative idea and carried it through to its complete practical application. His importance as the first entrepreneur in British telecommunication can be seen in his intuition and flair for programming; his capacity for improvisation and innate skill in turning obstacles into new opportunities; his ability to recognize his own limits and give priority to an entrepreneurial plan, rather than to individual interests.

His visionary power and capacity to programme in context emerge very clearly from the important documents he drew up, the 1836 business plan to present himself to possible backers and the contract-winning 1843 pamphlet on signal systems along railway lines.

The long-sightedness and lucidity of analysis of the business plan[3] are surprising for the times. Cooke foresaw governmental control of the telegraph in emergencies,[4] its possible use by merchants and businessmen[5] and ordinary people in order to keep alive

---

[2] "All or nothing" is the English translation of the motto adopted by the Electric in 1846. Frequently used in the crests of many English aristocratic families, it can be found on any document with the Electric's seal, like the shareholder registers. BT, *Electric Telegraph Company, TGA 1/6, Shareholder books 1846–1856*.

[3] The business plan has reached us thanks to an extensive extract made by his brother, in a posthumous publication. Cooke Thomas Fothergill, *Authorship of the Practical Electric Telegraph of Great Britain*, London, W.H. Smith and Son, 1867, pp. 96–111.

[4] "The Electro-Magnetic Telegraph gives great facility for the exercise of Government control in cases of emergency, without interfering with the public enjoyment of its fullest advantages on ordinary occasions. ... Where government despatches are most frequent, separate wires might be appropriated to their use. If, in this case, the public wires were made to pass through the Government office, Government could assume exclusive control, when, either their own purposes, or the avoiding of popular excitement, might render it necessary", Ibid., p. 102.

[5] "The advantages under this head [commercial affairs] are no less comprehensive and important. An immediate knowledge of the daily state of all the important markets, &c., would place the most distant cities of the kingdom on a footing in their mercantile transactions with the capital. The daily state of the money market so eagerly expected on the stock exchange of London, would be looked for with equal anxiety at Liverpool, Glasgow, and Newcastle, at the same moment. The capitalist of Glasgow might, without fear of an unfavourable change of price taking place in the interval, transmit to his agent in London orders for the immediate sale or purchase of stock or shares; and the banker of the country, when pressed in time of panic, might be preserved from stopping payment, by assurance from the capital of extended credit", Ibid., p. 102.

long-distance economic and personal relations.[6] Above all, however, he suggested introducing the telegraph network as a low-cost, highly remunerative infrastructure.[7] His intuitions on possible applications of electric communications included some concrete technical projects like protecting telegraph wires by inserting them into wooden channels or if necessary in steel tubes in the big cities.[8] Both solutions would be adopted more than 15 years later. Cooke did not even overlook finer points on marketing and self-promotion. He was quick to compare the electric telegraph to its optic rival, in order to make it more attractive to optic telegraph users, like the Government.[9] Furthermore, realizing that comparisons would be

---

[6] "The cases in which the convenience of individuals would be affected are innumerable; and perhaps there are few persons, however generally unconnected with the affairs of the busy world, who would not sometimes be spared, either a lengthened epistolary correspondence, or an expensive journey, by a few short communications through the Telegraph. The comfort of friends and relations, far distant from each other, would often materially involved, especially in cases of sickness", Ibid., p. 103.

[7] "Lastly, consider the sources whence a revenue, adequate to defray the original outlay, and current expenditure, of such an enterprise may be derived. If a Company undertake the execution, under the control of Government, a proportionate remuneration for the forwarding of official despatches may be looked for from that quarter. In each town where a Telegraph was established, various classes of subscribers would be admitted to the privilege of receiving and forwarding their private correspondence, and of being made immediately acquainted with the general news of the day.... In any interference, which an extensive establishment of the Telegraph might create with the Post Office, be viewed as an encroachment upon the revenues of Government, it will be remembered that communications by Telegraph may be taxed, as well as those by post; and as this interference would be almost insensible in any case, Government would gain, not lose, by the event", Ibid., p. 103.

[8] "The following plan for the construction of a conducting trough, is recommended by the Projector, as the cheapest and most durable, which has occurred to him; offering at the same time, an effectual mode of isolating the wires, and of protecting them from decomposition. He proposes that the wires be laid in separate channels, formed along the surface of narrow slips of baked wood, and covered in with a thin lath of the same material; and that this case be inserted into a trough of thick semicircular draining tiles, about thereof four inches in diameter, the space between the wood and tile being filled with water-proof cement. The whole may be guarded above by a flat tile, and buried from eighteen inches to two feet in the ground. Through large town, where the streets are frequently disturbed for the laying of gas or water pipes, iron pipes (the wires being isolated by wood), might be preferable", Ibid., p. 99.

[9] "The enormous expense of the present system is felt by Government in the short line between London and Portsmouth, the solitary instance in which it has been called into action, even in affairs of state. This expensive system labours under the following imperfections, which admit of no remedy.... The system here advanced, causing expense so small as to bear no comparison with that now incurred, and effectually supplying the deficiencies, and avoiding the imperfections above mentioned, offers advantages to the public, as well as to the Government, which it is hoped are not unworthy of consideration", Ibid., p. 98.

made at once between telegraph and postal correspondence, he knew how to anticipate and resolve the weak point of the lack of privacy in transmission by elaborating and introducing commercial ciphers,[10] another solution that would be adopted 15 years later.

What stands out in the 1842 pamphlet is Cooke's capacity to see and seize opportunities. Following the reports by the Select Committee on the Railways, the telegraph had received public approval as a signal system for the railways; it served to avoid the collisions and serious accidents which at their best blocked lines, and at their worst caused fatal accidents. With the excuse of illustrating the operations of the telegraph as a system of increasing railway safety,[11] Cooke elaborated an extraordinary document of promotion and advertising. The real purpose of his book was to show that a single-track railway was just as safe as a double-track one, but far more economical.[12] Up to then, two tracks had always been built for safety reasons – one for the outward journey and another for the inward one.

The two-track system was completely outdated by Cooke's proposal to run a telegraph line alongside the railroad in order to monitor the train's whereabouts. A control system operating in real time along portions of the track would guarantee not only safety but also punctuality.[13] In short, Cooke demonstrated that

---

[10] "Inscrutable secrecy is obviously of primary importance in the mass of telegraphic correspondence. This may, on a *prima facie* view, appear difficult to obtain, where the symbols are seen by a number of persons at the same time, without each corresponding party having a private signal book; but the accompanying 'Round Robin' cipher will at once show, from the innumerable changes, of which it is capable, that a published signal-book may be employed on the most private occasions. These changes depend upon the fresh relative arrangement of the circle containing the letters to that containing the figures, each time of the cipher being employed, however frequently it may be required during the day", Ibid., p. 101.

[11] "Two objects are sought in the following pages. The first is to add to the safety and efficiency of Railway communication, by means not more, but less, expensive than those now adopted", Cooke, *Telegraphic Railways*, p. 1.

[12] "The second is, to overcome some objections to the formation of auxiliary single Lines, by suggestions calculated to give them the safety and efficiency now supposed to be exclusively within the reach of double Lines", Ibid.

[13] "Should a train be delayed in its passage from one station to another, its non-appearance when due would lead to immediate enquiry, and as the stages would seldom exceed four miles in length, half that number would usually be the utmost distance that a messenger could have to go for assistance. The cause of delay would be immediately telegraphed to the divisions –stations, and if necessary, an engine or a train with carriages or workmen would be despatched from thence, or from the nearest depot, to the place of the accident. At present, a broken-down train sometimes waits helplessly for hours, till the next regular train arrives to assist it, – or to add to its difficulties, by a collision", Ibid., p. 24.

flanking the railway lines with the telegraph system would halve infrastructural costs, render more efficient the service and therefore returns, as well as increase safety. All-important was his insistence on the reduction of costs,[14] aimed at attracting the attention of the railway companies, which needed to reduce their enormous overheads. Cooke's message was taken on board by the railway companies, as can be seen by the escalation in telegraph lines installed alongside railroads between 1843 and 1845.

Cooke's capacity for improvisation and great belief in his project was shown particularly between 1836 and 1842, when the telegraph was ignored by those who should have recognized its enormous potential. But when faced by Faraday's refusal, Cooke had been quick to change tactics and separate the role of scientific partner from that of financial backer. Then, when the boards of both the London & Birmingham and Great Western refused to prolong his contract, Cooke decided to go it alone, bought the already constructed lines, and expanded the service by himself. He also showed much acumen, as in the way he introduced capital-saving technologies by replacing underground cables with overhead lines, a ploy that would prove essential in the development of the Electric, too.

His capacity to recognize his own limits emerged very clearly in the last years of the experimental phase, when, after the publication of the Pamphlet and his success with both Great Western and London-Blackwall, requests began to pour in from railway companies. Cooke recognized he needed financial partners bringing not only capital but also the business management know-how he lacked. And it was for this reason that he chose two men with a great technical and managerial experience like Bidder and Ricardo. Yet on this occasion

---

[14] "Where there is a very great traffic to require and justify the outlay, the double way will be preferable; in other cases, the single way will effect an immense reduction of expense", Ibid., p. 7; "If a Telegraph can be introduced into the system, its utility will stand upon the threefold basis of SAFETY, ECONOMY and EFFICIENCY; and, among the numerous telegraphic plans which Railway Companies have entertained, Electricity alone can hope thus to be the handmaid to Steam", Ibid., p. 11; "The consideration of expense has been stated to be one main point to which this essay is directed. It has been my object to show that Railways possessing every valuable property of those at present in use, and in some important respects their superiors, may be constructed and maintained at a great reduction of the usual outlay", Ibid., p. 28; "the Electric Telegraph may reduce the choice of double way, mixed way, or single way to a question of mere convenience, to be determined in each particular case by reference to the nature of the country and the amount of expected traffic", Ibid., p. 30.

too his faith in his entrepreneurial project was strong, for he did not give up all rights to his patents in return for immediate wealth, but used his quota to enter the Electric as a substantial shareholder and stayed there until the company was wound up.

While Cooke stands as the first entrepreneur in the field of tele-communications in Great Britain, Ricardo was the business brains behind the Electric. Once appointed chairman, Ricardo realized that Cooke's strategy of building and running signal systems along railway lines was too dependent on the lot of the railway sector. The company could only take off when it found a market where it could run a monopoly until others came in. According to Barton, Ricardo was the mastermind behind both the correspondence and intelligence services.[15] As a businessman and member of the Stock Exchange, he was only too aware of the importance of rapid and efficient information, and for this reason invested resources from his building concerns in order to construct the infrastructure and hierarchical organization required.

By the beginning of the 1850s, the take-off phase was over, thanks to the considerable capital coming from the railways. Ricardo had achieved his original aim and Electric was the first mover in the tele-graph communications market.[16] Anyone attempting to break into this new business would have to face the considerable entry barriers that Cooke and Ricardo had built up over the years, particularly the exclusive building rights that guaranteed many of the contracts conceding the railway wayleaves; the telegraph and line-component patents Cooke and Wheatstone had gradually deposited; Ricardo's contracts for supplying news items to the press; the existence of an infrastructure of about 2,000 miles of telegraph lines linking frequently and regularly the main British towns and cities.

Cooke and Ricardo's initial decisions left a very clear imprint on the following development of the market and its technical characteristics.

---

[15] Roger Neil Barton, "New Media. The Birth of Telegraphic News in Britain 1847–68", *Media History*, 16/4, 2010, pp. 379–406.

[16] This desire to create something new and get there first in a new marker is voiced by Ricardo himself: "The promoters [the Electric] had great difficulties to encounter. They had to meet the indifference or distrust of the railways, and the ignorance and apathy of the public. They had, in fact, to create a system, to invent a language and learn it themselves, and then instruct a staff", The Electric Telegraph Companies, *Morning Chronicle*, 2 June 1852.

2.2   The optic vs the electric telegraph

The electric telegraph was the first practical application of electricity and a first form of almost instantaneous long distance telecommunication, thereby constituting a new scientific paradigm. But though it was indeed extremely innovative, it was perforce subject to the inertia caused by present and past technologies and organizational systems which together with political and managerial decisions conditioned its development.[17]

The optic telegraph is generally considered the precursor of telecommunications. It consisted in coded messages sent long distance by means of sighting/transmission towers equipped with various signalling instruments which, according to period and place, could be long wooden arms, mirrors or other visual or acoustic appliances. Where governments had promoted and run the national construction of an optic telegraph network, the introduction of the electric telegraph was inevitably slowed down and encountered more difficulties. France, for example, was the most clamorous case, in that its optic telegraph service forced the Government to adopt for its electric telegraph system the Foy-Breguet apparatus, which reproduced exactly the movement of the mechanical arms of the telegraph towers. In this way, the personnel did not have to be retrained. But the result was slower and less efficient than the electric telegraph, and consequently found less favour and growth.[18] In the same way, in Spain, the delay in setting up an optic telegraph network until the late 1840s meant Government and Parliament were reluctant to replace it at once, so the installation of an electric telegraph was put off until a decade later.[19]

In Britain, there was an early interference between the two systems as far back as 1816. A certain Francis Ronalds, the inventor of a pioneering electric telegraph sent a letter to the Admiralty, which

---

[17] Joel Mokyr, "Technological Inertia in Economic History", *The Journal of Economic History*, 52/2, 1992, pp. 325–338.

[18] Louis Figuier, *Exposition et Histoire des principales découvertes scientifiques modernes*, Paris, Langlois et Leclerq-Victor Masson, 1851, pp. 147–162.

[19] Sebastián Olivé Roig, *Historia de la telegrafía óptica en España*, Madrid, Ministerio de Trasporte, Turismo y Comunicaciones, 1990; Sebastián Olivé Roig, *El nacimiento de la telecomunicación en España*, Madrid, Cuadernos de Historia de telecomunicaciones nº 4, Escuela Técnica Superior de Ingenieros de Telecomunicación, 2004.

was the proprietor and direct manager of the only optic telegraph line, the one between London and the naval port of Portsmouth,[20] in which he presented his invention and asked permission to install it. The Admiralty rejected the offer, saying it already possessed an efficient signalling system and saw no reason to change it.[21] Cooke insisted and stressed the indubitable advantages of the electric over the optic telegraph, carefully listing both the cost and the evident technical defects of the latter:

I.   The semaphore is only available during daylight.
II. Even by day it is interfered with in its action by partial rains, storms, morning and evening mists, and by the dense fogs of winter, for weeks together.
III. The utmost vigilance is constantly required at each station; and a degree of brevity, often objectionable, must be observed in the despatches.
IV. The symbols are publicly displayed, and their meaning is liable to detection.[22]

While the electric telegraph,

> causing expense so small as to bear no comparison with that now incurred, and effectually supplying the deficiencies, and avoiding the imperfections above mentioned, offers advantages to the public, as well as to the Government, which it is hoped are not unworthy of consideration. The Electro-Magnetic Telegraph ... is small, compact, and portable, yet very strong in its construction, and is unaffected in its action by darkness, or any obstacles of the atmosphere; the first signal being an alarum, which continues its call at intervals till answered, demands no vigilance.[23]

This argumentation was taken up by Wheatstone, who upheld it during his interview with the Select Committee, which was quoted by many newspapers.[24] Already in early 1841, when the telegraph

---

[20] Geoffrey Wilson, *The Old Telegraphs*, London, Phillimore, 1976, pp. 11–63.
[21] Jeffrey Kieve, *The Electric Telegraph, A Social and Economic History*, Devon, David and Charles Newton Abbot, 1973, p. 16.
[22] Thomas Fothergill Cooke, *Authorship of the Practical*, p. 98.
[23] Ibid.
[24] "Mr. Wheatstone states the advantages which the electric possesses over the ordinary telegraph as follows: – 'It will work day and night, but the ordinary telegraph will act only during the day. It will also work in all states of the weather, whilst the ordinary

service was still far off, it was already considered far superior to the optic version, to such a point that Joseph Hume asked the First Lord of the Admiralty about the suitability of replacing the old optic system between London and Portsmouth,[25] a task which Cooke was commissioned to carry out a few years later.[26]

In conclusion, the optic telegraph was never considered a serious rival to the electric version. Once the latter had been perfected, it was unanimously acclaimed the cheapest, most rapid and private form of communication. Great Britain's one optic line between London and Portsmouth was momentarily kept as a reminder of the need to link the fleet protecting coast and colonies with the central power. It was an imperative strategic need, which would later be replaced by the submarine cable and the Empire's Red Line, inaugurated in 1902. But that is a later story.

2.3 The pneumatic vs the electric telegraph

Unlike the optic telegraph, the pneumatic system was initially deemed a serious competitor to the electric version and at least in the initial phase seemed to be the winner.

---

telegraph can be worked only in fine weather. There are a great many days in the year, during which no communication can be given by the ordinary telegraph, and besides, a great many communications are stopped before they can be finished, on account of changes in the atmosphere; no inconveniences of this kind would attend the electrical telegraph. Another advantage is, that the expenses of the several stations is by no means comparable to that of the ordinary telegraph; no look-out men are required, and the apparatus may be worked in any room where there are persons to attend to it. There is another advantage which the electric possesses over the ordinary telegraph – viz., the rapidity with which the signals may de made to follow each other. Thirty signals may be made in a minute, a number which could not be made by the ordinary telegraph'", "Fifth Report of the Select Committee on Railways", *Yorkshire Gazette*, 22 August 1840; "Fifth Report of the Select Committee on Railways", *Preston Chronicle*, 22 August 1840.

[25] "Wished to know whether there was any intention on the part of the Government to adopt any new system of telegraph between London and Portsmouth. A new plan had been suggested, by which communication might be kept up day or night, with the greatest precision and rapidity", Joseph Hume's point of order to the First Secretary of the Admiralty, Richard More O'Ferrall, HC, Deb 5 March 1841, Vol. 56, cc. 1360–1361.

[26] The optic line was dismantled only after 1850. On 8 April 1850, Member of Parliament Thomas Thornley asked the First Sea Lord Sir Francis Baring if the optic telegraphs between London and Portsmouth were still in use and were of any use. The minister's answer was crystalline, given his reply: "the old system was done away with, and that the Admiralty were selling the old telegraphs as fast as they could", HC, Deb. 08 April 1850, Vol. 110, c. 66.

Probably because it was abandoned later on, pneumatic technology is little mentioned in the literature, even though it lived a period of great success in the mid-nineteenth century, when technicians and engineers tried to apply it both to driving force and long-distance communication.

Nowadays the pneumatic post is a communication system in which compressed air sends a capsule containing a hard-copy message at high speed through a tube. Originally, pneumatic signalling systems were similar, though varying in kind. For example, both the systems used by the Liverpool & Manchester and the London & Birmingham were based on a mechanism which activated a sound similar to a whistle by means of pressed air. When the convoys were ready to be hauled by the funicular system, the signaller moved a level which sent compressed air through a tube to the operator together with a whistling sound for the steam engine to be turned on.[27]

At the end of the 1830s, the pneumatic telegraph was twice preferred to the electric system, and as we have seen, Cooke paid the consequences. Pneumatic signal systems were simpler to construct, less expensive, but their trump card was that their messages were more intelligible. This was why in the early 1840s the press quoted the two above cases and saw in the pneumatic telegraph the future of distance communication, flanked by the electric telegraph in a minor role.[28] Corroboration came from other applications, in particular the

---

[27] "Electricity was thought of as a quicker signal agent, and some successful experiments were tried with it, but experience has proved that whistle is more advantageous and suitable at every respect", *Guide to the London and Birmingham Railway*, citation in Kieve, *The Electric Telegraph*, p. 28.

[28] "Notwithstanding all we have heard and seen of this Electrical Telegraph, (for we have seen a working model of it in London) we should prefer for convenience, simplicity, and, cheapness, the mode of conveying communications through the pneumatic or atmospheric tubes, which we recommended in *Mercury*, of July 6, 1834, in an editorial article entitled 'Rapid Conveyance of Intelligence – Pneumatic Tunnels, and Conjectures as to the attainable speed'", "Electric Telegraph on the Great Western Railway, and the Pneumatic Tube", *Liverpool Mercury*, 25 September 1840. "The electric telegraph is, no doubt, a highly ingenious and useful discovery, but it would appear to be extremely limited in its range of operations, two persons only being apparently able to hold communication at one time, and that communication of the briefest and least complicated kind. But, by the device I allude to, it is quite conceivable to transmit the whole mails and correspondence of the country, with almost any require rapidity, simply by a line of pipes or tubes laid from one place to another, above, or under, or along, the surface of the ground, having a stationary engine at one end to exhaust the air in front of an air-tight piston playing in the tube. ... On a large scale the plan is still

atmospheric railway, which was beginning to be exploited with a discreet success. Following the opening of a few brief lines in the late 1830s,[29] two more ambitious projects were set under way for the London, Brighton & South Coast Railway and Brunel's South Devon Railway Company.[30] Both started to function partly in 1846, both showed notable defects during the harsh snowy winter of 1846–1847 and the following summer. In particular, there was notable damage to the leather valves inside the air tubes and the animal fat used to prevent the valves from drying out attracted swarms of mice, all raising maintenance costs. Within a few years the atmospheric railway was dismissed as untrustworthy and too expensive and therefore abandoned, probably sealing forever the fate of the pneumatic telegraph.[31]

So despite the first successes and acclamations in the early 1840s, the pneumatic telegraph disappeared out of sight. As the electric telegraph became more and more efficient, especially over long distances, the pneumatic system developed greater technical difficulties and rising infrastructural costs. While it had proved more economical than the electric version on a small circuit, it encountered greater technical difficulties and higher costs on a larger scale.

The destiny of pneumatic technology in general was similar to that of the cable-hauled railway system with stationary steam engines. Between 1835 and 1845 railway systems using stationary steam engines rather than mobile ones had a moment of great success. While locomotives were noisy and polluting, the cable-hauled system had the advantage of being silent and polluting only where it was connected to a steam engine. Already partially used on the Liverpool & Manchester and London & Birmingham lines, it had its

---

in theory, but if successful in practice, it would give the great desideratum of a most rapid mode of sending information from one point to another without any unnecessary (so far as this single object goes) apparatus of locomotives, rails, stations, ground for these &c. &c, requiring nothing more than the mere tubes and the exhausting engine at the end", "An Auxiliary to the Penny Post", *Bradford Observer*, 6 June 1844.

[29] See in particular the *Dalkey Atmospheric Railway* and the *Compagnie du Chemin de fer de Paris à Saint-Germain*. Mallet, *Rapport sur le chemin de fer établi suivant le système atmosphérique de Kingstown à Dalkey, en Irlande*, Paris, Carilian-Goeury et V. Dalmont, 1844.

[30] R.A. Buchanan, "The Atmospheric Railway of I. K. Brunel", *Social Studies of Science*, 22/2, 1992, pp. 231–232; Charles Hadfield, *Atmospheric Railways*, Gloucester, Alan Sutton Publishing Limited, 1985; Arthur R. Nicholls, *The London & Portsmouth Direct Atmospheric Railway: "A Mere Puff of Wind"*, Stroud, Fonthill Media, 2013.

[31] Howard Clayton, *The Atmosphere Railways*, Lichfield, private publication, 1966.

endorsement in 1839, when it was adopted on the whole circuit of the London & Blackwall. However, alternatives to mobile locomotives were doomed to failure over the long period because of their technical difficulties and consequent hike in costs.

They may well have been victims of their own success, for as they went into use, the public took to them and so increased the need for improving communications over ever-extending distances. And at the same time, rival technologies like the electric telegraph and railways based on independent mobile engines were showing greater flexibility and grasping the opportunities beyond the reach of the other technologies as they went on enlarging their networks.

However, the pneumatic telegraph did not die out, but developed an interesting symbiosis with the electric telegraph. In the mid-1850s, the Electric began to construct and adopt a system of pneumatic post to link its various city offices, for pneumatic communications continued to offer notable advantages over short distances, and in fact went on being used for well over a century.[32]

2.4 Gas tubes and electric cables

The infrastructure for the distribution of gas for public illumination influenced in a radical way the development of urban telegraph networks, though being overlooked both by contemporaries and the literature. London was the first city to adopt a network of tubes for the gas lamp illumination of the streets, with the system progressively spreading to other British cities later, in the 1820s.[33]

During the initial heroic phase, the telegraph apparatus were linked together with naked copper wires and lacked any kind of insulating cover. Copper was a good conductor, the laboratories where the experiments took place were relatively dry, the distance between one telegraph and another not too long, so the method of constructing the telegraph lines was not originally considered to

---

[32] D.G. Clow, "Pneumatic Tube Communication Systems in London", *Newcomen Society for the Study of Engineering and Technology – Transactions*, 66, 1995, pp. 97–119.

[33] Janet Thomson, *The Scot Who Lit the World. The Story of William Murdoch, Inventor of Gas Lighting*, Glasgow, self-published, 2003; Leslie Tomory, *Progressive Enlightenment. The Origins of the Gaslight Industry, 1780–1820*, Cambridge (USA), MIT Press, 2012; Leslie Tomory, "Competition in the Early History of the London Gas Industry", *The London Journal*, forthcoming.

pose any serious technical problems. For example, though Cooke had already used impermeable cables for submarine experiments,[34] he still thought that economies could be made and overhead lines built with uncovered copper along the experimental lines of the Liverpool & Manchester.[35] As this system showed considerable dispersions of electricity, Cooke followed Stephenson's advice and inserted the copper wires within tar-covered wooden beams.[36] This system also proved inefficient, but what was worse, so expensive that the Liverpool & Manchester turned to the pneumatic alternative.

The first interaction between telegraph lines and gas distribution tubing took place some months later, along the Great Western Railway. Again Cooke attempted to build the telegraph wires Liverpool & Manchester style, but Brunel suggested an innovative technical solution, which was to insert isolated cables within steel tubes.[37]

---

[34] In May 1837, partly on Wheatstone's advice, Cooke had turned to the Enderbys, makers of impermeable cloths and cables used on the high seas. The Enderbys furnished Cooke with a waterproof cable he passed a copper wire through. But Cooke intended to use it for underwater experiments and not on overland line... "They appear very friendly to me. I am going down to Greenwich to-morrow morning to see their works, and arrange a method of covering our wires with yarn and include them in a rope, for our cross-Thames experiment. I sent the wire off yesterday. This rope, 1,500 feet in length, including 6,000 feet of wire, is to be ready by the close of the week", IET, WFC, SC, Mss 007, "Letter to W. F. Cooke's Mother", 23 May 1837.

[35] "By strenuous exertions I succeeded in collecting the above vast quantity of wire, cleared the huge workshop of men and lumber, by the constant labour of from 30 to 40 men, and had nearly half a mile of wire arranged by Friday night; proceeding slowly on Saturday morning, having to teach all the men employed – viz., 8 carpenters, 2 wire-workers, and 8 boys – their distinct duties, we got forward more rapidly towards evening, and at 5 o'clock, when the men left off work, I had about four miles of wire well arranged, and hope to get all nearly done by to-morrow night. You imagine the task when I tell you that 2,888 nails have been put up for the suspension of the wires", IET, WFC, SC, Mss 007, "Letter to W. F. Cooke's Mother", 2 July 1837.

[36] "He [Mr. Stephenson] has allowed us to go to another £100 expense on account of the Company, and only added as a condition that all should be perfect, for 'I do not like to be laughed at for a failure'. He said 'We will have not only the Directors, but half the nobility and everybody else to see it; only let all be sure;' and I sparing no pains to comply with his injunctions. Making out drawings, forming plans, obtaining materials, have occupied my day pretty busily, and I am likely to have a week similarly employed. I have just given orders for 5,000 feet of wood to be sawn in a particular manner, with grooves for the wires, which I am going to have boiled in coal tar previously to laying down. Our wire is all ready", IET, WFC, SC, Mss 007, "Letter to W. F. Cooke's Mother", 25 July 1837.

[37] "Orders are already going round to all my workmen to re-commence operations most accurately, and I think it possible three miles may be done with fine weather by the end of the next week, but I may have to wait for timber, as Mr. Brunel seems inclined

Since Brunel was an expert on urban infrastructures, he must have been thinking of gas distribution networks, and as if to prove the point steel tubes were commissioned from a company which made urban mains.[38] Using steel tubes for telegraph lines proved extremely expensive, though the good insulation, guaranteed by cotton and natural rubber wrapped around the copper wires, convinced Cooke to repeat the operation along the London & Blackwall.[39]

Meanwhile, at the beginning of the 1840s Cooke, by then proprietor of the Great Western telegraph network, had installed in it the first overhead lines constructed on the basis of his patent.[40] These lines, composed of naked galvanized wires on insulators held up by wooden poles, were considerably cheaper than those using steel tubes. From then on, practically all the lines commissioned from Cooke were constructed with overhead lines running alongside the railway tracks. All these lines, installed pre the Electric, can be classed as extra-urban, in that all telegraph offices were then located in railway stations, and therefore in areas which were still relatively peripheral. The problem of how to build telegraph wires in urban contexts was resolved by legislation, when parliamentary authorization was granted to the Electric, and reference was once again implicitly made to gas tubes. Article 35 of the Act read, in fact,

---

to change his plans. This I shall know on showing my plans to-morrow morning", IET, WFC, SC, Mss 007, "Letter to W. F. Cooke's Mother", 28 May 1838. "Besides, the plan is a novel one which Brunel has decided upon, and I have thought less upon it than on any other. I mean to lay down about a quarter of a mile in the garden over and over again, till each man knows his duty. The materials will do again on the main line, except the wires, which will have their jackets worn out; and even they can be burnt, cleaned, and re-covered. It will only cost me £4 or £5, and the expense to Drayton is £3,159", IET, WFC, SC, Mss 007, "Letter to W. F. Cooke's Mother", 31 May 1838. "I rode nearly 30 miles yesterday, giving final orders for the delivery of goods, and making of small pieces of machinery, &c., &c., all of which required thought and calculation. Our plans are entirely changed; and I am to lay down the wires in iron pipes. In consequence, all the instruments and tools before made are useless; and one fresh set is nearly complete. The plan now adopted is the most perfect I can desire, but the cost is always feared would be the objection", IET, WFC, SC, Mss 007, "Letter to W. F. Cooke's Mother", 8 June 1838.

[38] The firm was John Russell, Gas Tube Manufacture & c. from Wednesbury, Francis Celoria, "Early Victorian Telegraphs in London's Topography, History and Archaeology", in *Collectanea Londinesia*, being Special Paper N° 2 of the London and Middlesex Archaeology Society, 1978, p. 425.

[39] Ibid.

[40] *Improvements in Apparatus for transmitting electricity between distant places*, Patent n° 9465 (England), 8 September 1842.

And be enacted, that it shall be lawful for the Company from Time to Time to lay down and place under any Street, and either along or across such Street, but not above the same, any Pipes or Tubes, not being of larger Size than Three Inches Bore, which shall or may be necessary or convenient for the Purposes of any Electric Telegraph or intended Electric Telegraph, or for conveying the conducting and other Wires of such Telegraph along or across such Street.[41]

The regulation clearly aimed at curbing the increasing congestion in the streets of London and other big cities. Thanks to Cooke's past experiments, the Electric used right from the beginning earthenware tubes, which were not only cheaper than steel, but were not subject to electric dispersion.[42]

Given the density of urbanization and the fact that the Act authorizing the Electric was the only provision made for telegraphs, the apparently marginal parliamentary decision was destined to have two important consequences: a drastic hike in infrastructural costs and an increase in cable technology.

The first consequence had already been revealed by the press, which had highlighted how British telegraphs were more expensive than those in the German areas, and far more than in the States. The obligation to build underground lines in urban areas not only forced the Electric to offer a service at higher prices than originally estimated, but also made the entry of new competitors more complicated. Nevertheless, while the price hike was seen negatively, the increase in cable technology led to extremely positive outcomes. The need to install urban lines within tubes obliged telegraph engineers to look for technical solutions to prevent the underground humidity from generating electricity dispersion and hampering communications. Continual improvements were made to the natural rubber coatings of the copper wires and progress in this field would be such an advantage for British submarine telegraphs, which forged the way in the international market.

2.5   The cumbersome presence of the Penny Post

The post and the telegraph were like two sibling services where the elder ate his fill and left the younger one just enough crumbs for

---

[41] Article XXXV, *An Act for forming and regulating "The Electric Telegraph Company", and to enable the said Company to work certain Letters Patents* (18 June 1846), 8 Victoria, c. XLIV.

[42] Ibid.

survival. Both based their existence on the international exchange of information, but the post, the first to arrive on the market and state-run to boot, offered a universal service while the telegraph inevitably specialized in a rapid service, which was efficient but expensive and therefore select and exclusive.

In 1839, while the telegraph was going through its experimental phase, Rowland Hill designed and directed the introduction of the Penny Post,[43] which consisted in establishing a flat charge for all letters sent within the country, independent of the distance covered,[44] guaranteeing the sender the possibility of pre-payment, thanks to the franking system. The post, which up to then had been expensive and accessible only to the wealthier classes, went through a rapid democratization, both in terms of quantity and quality. Quantitatively, there was an undeniable increase in users and postal traffic, while qualitatively there developed a progressive differentiation in the users, i.e., the universalization of the service. To increase the access of the new social classes were also some radical political campaigns, like the anti-Corn League's, which used stamped letters for their propaganda.[45] Another sign of the democratization of the postal service was the considerable increase in the sale of products for correspondence, like portable writing desks, letter paper, various kinds of inks and pens.[46]

The telegraph was moving its first steps, from experimenting and then offering a public service right in the years of full affirmation and expansion of the Penny Post. The low cost of postal correspondence saturated the market and probably prevented the promoters of the telegraph service from seeing their service in universal terms. Though, for example, Cooke had foreseen a correspondence service in his business plan, he had initially developed the telegraph as a railway signal system. When Ricardo became chairman of the Electric and began to plan his correspondence service, he saw it at once as targeted at those who for business and commerce needed rapid communications. In autumn 1847, the Electric advertised its service in the following terms:

---

[43] Frank Staff, *The Penny Post: 1680–1918*, London, Lutterworth Press, 1964.
[44] R.H. Coase, "Rowland and the Penny Post", *Economica*, 6/24, November 1939, pp. 423–435.
[45] Catherine J. Golden, *Posting it. The Victorian Revolution in Letter Writing*, Gainesville, University Press of Florida, 2009, pp. 43–82.
[46] Ibid., pp. 135–142.

The Electric Telegraph Company has just issued a prospectus of the manner in which they intend establishing the "Commercial Telegraph" for the exclusive use of mercantile men in London and the chief towns in the Kingdom, distinct from the telegraph reserved for the use of railways and the Government. ... At the central station in London, at the back of the Bank, is to be opened a subscription-room. Like Lloyd's, and the subscribers are to have the right of entrance to all the stations of the company throughout the country, so that they will be enabled to communicate hourly with London, if necessary, and obtain instant information of what is occurring at their places of business, as well as the sixty chief towns in the kingdom with which the telegraph, at its opening, will be in connection. ... But when a merchant can make instant communication from some central station with such places as Dover, Liverpool, Edinburgh, Birmingham, Bristol or Manchester, or with his correspondents in those places as to time, he can receive intelligence from them all within a quarter of an hour, we think that a material change will be effected in the present mode of conducting mercantile correspondence.[47]

Some years later, in a famous letter written in defence of the company, Ricardo repeated the concept and affirmed:

The telegraphic system is designed for important and urgent messages ... A long and expensive journey is prevented, or a necessary one hastened – a bill accepted or protested – remittance made – a purchase effected or countermanded – an important witness is summoned – the arrival or the loss of a ship is announced – an insurance is effected – advice is asked, or orders given – in fact, an endless variety of important announcements, questions, and replies pass daily through our numerous receiving-houses, and, I can honestly aver, have for the most part effected, and have been acknowledged to have effected, an economy and convenience far beyond the 1d. per mile of our tariff.[48]

To sum up, the universal correspondence market was also open to the telegraph, which could have become complementary to the post. However, because of its low prices and high social and geographic presence, the Penny Post was an unreachable rival over a short period, partly because it was run directly by the Government. In a very frequent dynamic between old and new media, the first to arrive forced the late-comer to carve itself out a niche in order to

---

[47] "Electric Telegraph Company", *Leicester Journal*, 24 September 1847.
[48] "The Electric Telegraph Company", *Leeds Intelligencer*, 12 May 1849.

avoid direct competition.[49] The relation with the post explains why the telegraph was seen right from the beginning both by its operators and the public as an exclusive service reserved for big business and commerce.[50]

2.6   The symbiosis between railway and telegraph

There was a reciprocal interdependence between the railways and the telegraph, which guaranteed both of them a vigorous development, but at the same time created moments of attrition. The symbiosis between the two went through three distinct phases: the experimental introduction of the telegraph, thanks to the consensus and capital of the railway companies in the 1830s; the telegraph generally adopted as the signal system along the railways, after the Select Committee Report and the publication of Cooke's pamphlet in the early 1840s; the injection of capital and know-how into the Electric from 1848 to 1850.

During the first phase two factors pushed the telegraph into the arms of the first railway companies: their need for a signalling system for convoys moving along tracks and the telegraph's need for initial capital, easy to obtain from the railways, which were living

---

[49] On the birth of a new media: Roger Silverstone, "What's New about New Media?" *New Media and Society*, 1, 1999, pp. 10–12; Lisa Gitelman, *Always Already New: Media, History and the Data of Culture*, Cambridge, MA, MIT Press, 2006; Lisa Gitelman and Geoffrey B. Pingree (eds), *New Media, 1740–1915*, Cambridge, MA, MIT Press, 2003; David W. Park and Nicholas W. Janowsky, Steve Jones (eds), *The Long History of New Media: Technology, Historiography and Newness in Context*, New York, Peter Lang, 2011. For a wider interpretation of all the technologies, see Trevor Pinch and Wiebe E. Bijker, "The Social Construction of Facts and Artefacts: Or How the Sociology of Science and the Sociology of Technology Might Benefit Each Other", *Social Studies of Science*, 14/3, 1984, pp. 419–424.

[50] "It appears that the prevailing subjects of the dispatches vary according to the station from or to which they are sent. Thus, as might naturally be expected, in large commercial marts, such as Liverpool and Glasgow, they are chiefly engrossed by messages of mercantile firms and Business. Their prevailing subjects also vary much with the season of the year. Thus, in summer, the messages of tradesmen are greatly multiplied in consequence of the number transmitted by dealers in perishable articles, such as fish, fruit, &c. which must be supplied in regulated quantities with the greatest promptitude.... Independently of the direct use made of the electric telegraph by the general public, for the transmission of private despatches, the several companies have established, in various principal places, news rooms, where intelligence is from hour to hour posted, as it arrives from all parts of the World", Dionysus Lardner, *The Electric Telegraph Popularised*, London, Walton and Maberly, 1855, p. 76, 85.

their first boom.[51] As we have seen, Cooke recognized these needs and turned to the railway men, who gave him the possibility and money he needed. However, the experimentation was not only an advantage for the telegraph: the entry of two first-class railway engineers into line construction (Stephenson followed by Brunel) caused on two occasions an upsurge in costs, so forcing the companies that had backed the telegraph system to abandon it. Obliging Cooke to insert the telegraph wires in tubes delayed the spread of the telegraph, which took off when in 1842 Cooke used his own funds to set up overhead lines and thus slash installation costs.

During the second phase the telegraph played a greater part. The Select Committee recommended its use to the railways companies, a fact which was much publicized by the press, and in addition Cooke's pamphlet had shown that installation costs would be halved by passing from a double to a single track. A second railway boom followed, which made available the large amount of capital necessary for the development of the telegraph network. This virtuous circle, in which the expansion of the railways and the telegraph fed each other, generated new elements of interdependence. First of all, on an economic level, the activity of distance communication integrated with the transport system: the telegraph linked demand with offer, the railways perfected the correct allocation of the goods.[52] In the second place, the construction of telegraph lines alongside the railroads entailed the concession of wayleaves to Cooke and later to the Electric, in exchange for the labour of installation and maintenance. When competitors appeared, these contracts were implicitly

---

[51] E.F. Clark, *Going Public. The Formation of the Electric Telegraph Company*, in Frank A. J. L. James (ed.), *Semaphore to Short Waves*, Proceedings of a conference on the technology and impact of Early Telecommunications held at the Royal Society for the encouragement of Arts, Manufactures and Commerce on Monday 29 July 1996, organised by the British Society for the History of Science and the RSA, RSA, London, 1998, p. 27; Geoffrey Hubbard, *Cooke and Wheatstone and the Invention of the Electric Telegraph*, London, Routledge & Kegan, 1965, pp. 113–117.

[52] James Beniger, *The Control Revolution. Technological and Economic Origins of the Information Society*, Cambridge MA, Harvard University Press, 1989, pp. 219–290. The economic symbiosis between the two services is well exemplified by some anecdotes spread by the press, according to which some railways passengers who had lost an umbrella or even a baby, were able to get it back, thanks to the combined effect of the telegraph and the train. The telegraph allowed the lost object to be identified, while the train allowed it to be physically returned to its owner. *The London Anecdotes for all the readers*, London, David Bogue, 1848.

transformed into exclusive rights for the Electric. Thirdly, the introduction of telegraph signal systems made rail transport more efficient (fewer delays, growth in traffic thanks to timetables, and a speedier resolution of accidents and breakdowns) to such a point that railway companies could not do without them. The alternative was higher fixed and rolling costs and being priced out of the market.

These new ties of interdependence explain in their turn how the third phase came about. During 1848 the Electric's troubles were very clear and much discussed in the press. The need for new capital to bring down initial costs united with a need for technical competences leading to a rationalization of the new correspondence service persuaded Ricardo to involve new shareholders closely associated with the railway companies. The latter, considering the ties created in the second phase, could not allow the Electric to go bankrupt,[53] so the railway companies became directly involved in the Electric by means of a series of crossed shareholdings, which fully involved the railwaymen in the holdings and board of directors.

2.7   Conclusion

Our detailed analysis of the events of the first 15 years of British telegraphs and study of its interactive relation with other technologies, allows us to identify at least six constitutive choices: the telegraph service divided into two activities, managing the signalling service for the railways and transmitting correspondence and news; telegraph correspondence was reserved for a commercial and financial elite which needed urgent communications; the Government had its private line to guarantee national defence (London-Portsmouth); the telegraph network was mainly composed of less expensive extra-urban networks which guaranteed long-distance communication, an impossibility for other technologies like the pneumatic telegraph; urban lines had to be put underground, and therefore the technology for lagging the cables had to be improved; railway capital was directly involved in the management of the only telegraph company.

---

[53] Contemporaries were also well aware of the symbiosis: "in the Management of railway Business in all countries, but more especially upon our own ever crowded and over worked lines, the telegraph has become an indispensable accessory, without which this mode of locomotion would be deprived not only of its efficiency but its safety", Lardner, *The Electric Telegraph Popularised*, p. 78.

Together these choices determined the placing of the telegraph in various reference markets: the construction and running of signals (completely new); correspondence and circulation of intelligence (already existing but below strength); government communication for military/strategic reason (already existing but previously taken up by the optic telegraph).

The placing of the Electric in these markets determined the birth of three main categories of users: railway companies, the press and businessmen. They were flanked by the Government, which played an essential but quantitatively minor role.

The reduced dimensions of the market, caused by the exclusive nature of the service, the high cost of the urban lines and the direct involvement of the railway companies in the capital stock favoured the consolidation of a monopoly and consequently, the strengthening of the entry barriers erected by Cooke and Ricardo. These features would condition over the brief and medium length periods the entry of new competitors. The Government's interest for strategic reasons and the investment in waterproofing the cables would emerge over the long period as triggering the development of British submarine telegraphs.

3
From Monopoly to Free Competition

The early 1850s were a key period in the development of the tele-graph correspondence market, with the passage from the Electric's monopoly (1850) to the existence of four companies (1855). New arrivals were always interpreted as a direct challenge to the Electric, by the company itself, the often-hostile press and complaining clients. Following the classic strategy of latecomers, the new tele-graph companies tried to enter the market by inserting themselves in niches left vacant by the Electric, so they exploited the thousands of miles of railway devoid of telegraph systems and the scores of impor-tant industrial and commercial centres still lacking coverage. The new entries had extremely different approaches, for while some chal-lenged the Electric with marketing and legal duels, others moved the battle over to technological innovations and ensuing efficiencies. On its part, the Electric fought the war in defence of its monopoly, attacking on all sides (political and technological) and reaching the end of the five-year period with a telegraph network which was almost twice as big as all the others put together.

3.1 The Electric against all

During 1849 the press took advantage of the Electric's troubles and launched a virulent campaign against it, particularly its monopoly.[1]

---

[1] "We deeply regret to be obliged to stigmatise the manner in which the Electric Telegraph Company's officers carry on the business connected with this wonderful

Initially it was accused of inefficiency, delays and high prices.[2] Later, in a Rossini-like crescendo, the charges rained down on Ricardo, who was accused of manipulating the intelligence market.[3] Despite his declared position as a liberal and defender of free trade, the press accused him of blocking the entry of new competitors.[4] Openly

---

and highly important invention. To advertise 'instantaneous communication' and yet take 6½ hours to transmit a message of sixty words is really too bad. It is high time that something should be done to curtail the efforts of these avaricious monopolists, and that the enormous charges which they make should be brought to a reasonable and fair standard. As matters at present exist, the public are outrageously taxed; their business, in many instances, impeded; and the utility and value of the Electric Telegraph itself, in a great degree, unappreciated", "The Americas", *Glasgow Herald*, 9 April 1849.

[2] "Electric Telegraphing in England advertised to be instantaneous, accurate, and cheap", *London Standard*, 28 April 1849; "The numerous cases of gross neglect and inattention which are continually occurring in the management of the Electric Telegraph Company, call for reprehension in the strongest terms. This company, avowedly established for the convenience of the public, the facilitation of business, and quick transmission of intelligence, has departed from all these professions in more than one instance. Not only, however, has the public to complain of manifold delays and inaccuracies, but also of the exorbitant rates charged for the transmission of messages. 'Carpe Diem' appears to be the motto of the company, and whilst there is no competition, the disposition to act up to it is fully developed. Such a state of affairs cannot, however, be suffered in a commercial community", *Newcastle Journal*, 2 June 1849; "More misdoing of the Electric Telegraph", *London Daily News*, 6 October 1849.

[3] "Certainly it was a legislative error which gave to a body of men the right to speculate with so very potent an instrument. Nothing approaching to the character of a monopoly should have been associated with any channel of public communication; but, in the case of the electric telegraph, there were circumstances which called for the exercise of more than ordinary caution. Communications sent through the Post Office, being written and enclosed, are by a form of secrecy protected, since an outrage must be committed before the contents can be ascertained; the seal must be broken before the letter can be read; but the message sent through the electric telegraph is obliged to be openly delivered. It is true that the meaning can be disguised by employing symbols, but in many instances unexpected events do not permit of any preconcerted arrangements. Moreover, secrecy becomes more and more difficult exactly in proportion as it is likely to be valuable... Great facts have strange facilities of being recognised; and while new grant to the fullest extent every legitimate right that appertains to individual enterprise, we cannot without dread contemplate a mode of communication which is established under conditions that possibly might enable a small corporation to appropriate all the advantages to be derived from priority of information.... It is not a fraud but a conspiracy, which has been exposed; and it must be crushed before it is matured, or no man can foretell the consequence. The electric telegraph is the source of information; the journals are the means of its distribution. Let the company establish the influence it evidently seeks to exercise over the public press, and any man capable of a second idea will perceive how vast a power would thereby be created. With the capability to torture or withhold the facts by which opinion is guided, and fears or passions of the populace excited, no Government could be safe, and commerce would be annihilated", *Morning Post*, 12 October 1849.

[4] "It may truly be said of J.L. Ricardo, Esq., M. P., chairman of the Electric Telegraph Company, that he is *sui appeteus* and *alieni profusus*. As one of the heroes of Free Trade,

hostile to the hold Electric was tightening in the intelligence sector, the newspapers had been enthusiastic about the entry of new companies, right from the first rumours of their impending arrival.[5] They felt that breaking Electric's monopoly would generate an immediate fall in tariffs, an improvement in the quality of communications and a greater freedom in the intelligence market.[6]

It is not surprising, therefore, that all through the early 1850s, the press organized detailed campaigns in favour of any competitors presented to the public and systematically against the Electric, at times at a level of slander.[7] It began in 1850, when Parliament granted the authorization for the establishment of the British Electric Telegraph

---

he was used as the cat's paw to move for the committee to inquire into the Navigation Laws with a view to their repeal – to remove the great national monopoly, and, in the blessed spirit of philanthropy, to ruin thousands by giving the untaxed foreigner the privileges of an Englishman. But this liberal-minded gentleman plays a different game where his own interests are concerned, and repudiates his own cry of cheap living – witness the exorbitant charges and exclusive system of the Electric Telegraph. It is time that public attention should be called to his glaring inconsistency, and to the fact that the said Electric Telegraph Company has hitherto evaded the charge legally due on it, of payment to the poor-rates of this kingdom in every parish through which their wires pass, by presuming on the accommodation afforded them by the railway companies in the use of their lines", "Consistency, to the editor of the Morning Post", *Morning Post*, 19 May 1849.

[5] "In one of your late numbers you have adverted with much force and justice to the anomalous fact that the electric telegraph, the most wonderful discovery of modern times, although in full practical operation in England, has remained hitherto almost without utility, and has served for little more than to excite 'wonder at the marvellous feats achieved by modern science'. As a commercial speculation, as you observe, it has been a failure. The public, whether for commercial or social purposes, have availed themselves of its vast powers to a most insignificant extent", "The Electric Telegraph to the editor of the Times", *Freeman's Journal*, 15 October 1850.

[6] "Application is intended to be made to Parliament in the ensuing session for Acts to incorporate four different telegraphic companies – viz., one for the establishment of a submarine telegraph between England and Ireland; another for a similar undertaking between England and France; the third is of a more general character, and, under the title of the European and American Printing Telegraph Company, comprehends a wide geographical range of action in the old and new world. These three companies propose to use the printing telegraph invented by Mr. Jacob Brett, while the fourth, under the denomination of the Magneto-Electric Telegraph Company, confines itself to the patents granted to W. T. Henley, and D. G. Foster, embodying the application of Faraday's beautiful discovery to the movement of magnetic needles, &c. It is rather difficult to state the precise sphere of action contemplated by this company, as the proposed incorporation is for the purchase and use of certain patents in Great Britain and Ireland – and elsewhere. With so much competition before us, it will go rather hard if we do not have a cheap telegraph presently", "The Electric Telegraph", *Royal Cornwall Gazette*, 6 December 1850.

[7] In a letter to the *Morning Post*, Fourdrinier, secretary to the Electric, specified that the line with exorbitant prices a journalist had complained about a few days earlier did not

Company[8] (henceforth the British), a company which operated with the telegraph patented by the Highton brothers and exploited railways lacking telegraph lines, most particularly in the North and in Ireland.[9] In the same way, the next year the papers backed the entry of the Magnetic Telegraph Company (the Magnetic), which presented itself to the public with a similar plan to that of the British (a new telegraph and coverage of territory not served by the Electric), but far less aggressive and invasive in manner.[10] Less in the public eye, but

---

belong to his company. The incident goes to show that at times criticism of the Electric passed the limits of objectivity and correctness. "The Electric Telegraph Company", *Morning Post*, 14 May 1851.

[8] *An Act for forming and regulating the British Electric Telegraph Company, and enable the said Company to work certain Letters Patent* [29 July 1850], 13–14 Victoria, c. lcccvi.

[9] "Their [of the British] first enterprise is a great line from London by Manchester to Glasgow, thence by submarine telegraph to the Irish coast, and then through Belfast and Dublin to Cork. Other two lines are those from London to the west and the south–east. In the infancy of the submarine telegraphs the proposition to cross the Irish Sea as its northern and narrowest strait is well deserving of approval, and the apparent circuitousness of the route is of no real importance. The line selected will afford the means of communication to a number of the chief cities of England, Scotland, and Ireland, and thereby it is likely to obtain a good traffic. The prospects of immediately carrying on this line are represented to be very good", *Essex Standard*, 10 January 1851; "Progress of the telegraph", *Glasgow Herald*, 27 January 1851; "Evils of Telegraphic Monopoly", *Leeds Mercury*, 1 February 1851; "The British Electric Telegraph Company", *Gloucester Journal*, 1 February 1851; "The British Electric Telegraph Company", *Worcestershire Chronicle*, 23 April 1851; "Competition, it is said, is the soul of business, and the British Electric Telegraph Company are about practically to test the truth of the adage. We understand that offices well adapted for their operations have been taken by the company in Manchester, and they are in treaty for suitable premises in Liverpool also. It is anticipated that the line will be open for the transmission of messages between Liverpool and Manchester in from two to three months from the present time, while the directors are on the eve of concluding arrangements with several railway companies which will enable them to commence, without delay, their works from London to Liverpool. It is the expressed intention of the company not only to introduce a lower system of charges, but the adoption of instrumental improvements and inventions", "British Electric Telegraph Company", *Manchester Courier and Lancashire General Advertiser*, 5 July 1851.

[10] "Upwards of 200 miles of insulated wire have already been laid down by the Magnetic Telegraph Company upon a portion of the Lancashire and Yorkshire Railway; and the new line of Telegraph between Manchester and Liverpool, belonging to his company, will be opened in a week or two. The Magnetic Telegraph appears to have fulfilled all the favourable anticipations that have been formed respecting it, wherever it has been employed", "The Magnetic Telegraph", *Durham County Advertiser*, 21 November 1851; "It is with pleasure we see that the English and Irish Magnetic Telegraph Company are coming before the public, already having quietly constructed, or made arrangements for the completion of their extensive system of telegraphs, which will at last, we think, afford what we have long needed – a sufficient and certain means of communication. Their main lines being laid underground will obviate that uncertainty which has hitherto been experienced in the working of the overground system in wet weather. The

nonetheless well backed, was the entry of the European & American Electric Printing Telegraph Company (the European), a subsidiary of the French Submarine Telegraph Company between England and France[11] (the Submarine), and the United Kingdom Electric Telegraph Company[12] (the United), which was inactive until the 1860s. There was also the Universal Electric Telegraph Company,[13] which never became operative.

The unconditional backing of most of the press for the new companies provoked a harsh reaction from the Electric,[14] which put out a series of particularly aggressive measures in defence of its dominant position on the market. For example, in February 1850 Ricardo used his position as a member of Parliament to oppose a second reading of the Bill for the British, declaring that it was against the interests of the Electric, but the house voted against him.[15] Some months later, the Electric lowered its out-of-town tariffs[16] and introduced particularly economical prices for communications between London offices.[17] When faced by the arrival of other competitors in mid-1851, the Electric repeated its strategy and lowered its prices

---

London, Manchester, and Liverpool line of this Company will be of the most approved construction; and is estimated, we think fairly, to be able to meet the whole of the requirements of the public between these the three largest towns of England. A large portion of the lines of the Company in the United Kingdom is already completed", "English and Irish Magnetic Telegraph Company", *Leamington Spa Courier*, 1 January 1853.

[11] "Parliamentary Applications for the Ensuing Session", *London Daily News*, 2 January 1851.

[12] "Telegraph Competition", *Newcastle Journal*, 7 May 1853; "United Kingdom Electric Telegraph Company", *Caledonian Mercury*, 23 May 1853.

[13] "The Universal Telegraph Company", *Hereford Times*, 19 November 1853; "The Universal Electric Telegraph Company", *Dundee, Perth, and Cupar Advertiser*, 9 December 1853.

[14] See one of the rare articles in its favour, "Electric: Telegraph Companies", *Leeds Intelligencer*, 15 March 1851.

[15] "House of Commons", *London Daily News*, 23 February 1850.

[16] "The Electric Telegraph Company has just issued a notice intimating that after Monday week the maximum charge for every twenty words for any distance, is to be ten shillings. The effect of this alteration will be a considerable saving to the public on all messages transmitted upwards of 280 miles, which comprehends communications between Glasgow, Edinburgh, Newcastle, and London and Liverpool", "The Electric Telegraph Company", *Bradford Observer*, 14 March 1850.

[17] "The Electric Telegraph Company have now made arrangements for the despatch of messages between their London stations, at a charge of one shilling for communication not exceeding 20 words. This new feature in telegraphing is likely to prove of considerable utility to the metropolitan public; and the vast extent of our modern Babylon

again, justifying itself with the improved insulation of the lines.[18] At the same time it took to court numerous inventors partly linked to the new companies, accusing them of infringing its patents.[19] It also convinced all the railway companies whose lines were linked with London to sign exclusive agreements, in order to prevent the British from linking its offices with the capital. Thus the British was forced to undertake a parliamentary battle to avoid having to construct its lines along highways and finance expensive underground cables.[20]

3.2   The British: litigation and marketing

During the early 1850s only four of the companies to receive authorization started to construct lines in open competition with the Electric: the British, the Magnetic, the European, the Electric Telegraph Company of Ireland (the Ireland). The British, the Magnetic and the Ireland all aimed at Northern and Irish territories, where there were thousands of railway miles unequipped with telegraph systems. The Submarine had anticipated the Electric in laying the first submarine cable between Dover and Calais, with the European as its land extension to London. However, both competitive models had initial disadvantages. Scotland and Ireland were decidedly not so remunerative, given their huge non-urbanized swathes and decidedly higher costs in laying and maintaining a submarine cable.

Despite a forceful publicity campaign, parliamentary authorization[21] and an attempt to lay a submarine cable between Ireland

---

renders such a speedy mode of communication between distant parts as necessary as a three penny omnibus", *Chelmsford Chronicle*, 23 April 1850.

[18] "We understand that the directors of the Electric Telegraph Company have received the report of their engineer to the effect that the re-insulation of their lines upon the new principle, patented by Mr. Edwin Clark, is completed, and that they have determined forthwith to make a considerable reduction in their charges for the transmission of the messages of the public, and at the same time to simplify their tariff, by making it as far as possible uniform", "The Electric Telegraph Company", *London Standard*, 24 September 1851; "Reduction of telegraphic charges", *Leeds Mercury*, 27 September 1851; "The Electric Telegraph Company", *Newcastle and Tyne Mercury*, 4 October 1851.

[19] BT, *Electric Telegraph Company*, TGA 2/1, *Half yearly report June 1851*.

[20] "Parliamentary Committees. The Electric Telegraph Companies", *London Daily News*, 20 May 1852.

[21] "Electric Telegraph Company of Ireland", *London Standard*, 28 December 1852.

and Scotland,[22] the Ireland went under in 1856, without ever managing to transmit a telegram.[23] The other three all started business before 1855, but adopted very different strategies. While the British focused on an aggressive marketing and a legal battle with the Electric, the Magnetic and the European invested in technological innovation.

Chaired by James Simpson, a well-known hydraulic engineer,[24] and managed by George Saward, in that period an unknown telegraphic manager, the British presented itself to the public as self-assured and unconventional. First of all, it offered the same range of services as the Electric: the construction of telegraph lines for railway companies, Government and privates; the maintenance and running of telegraph networks for third parties; the sale of licences for the exercise of the telegraph service with patents in hand; the construction of a business network for correspondence and the circulation of intelligence.[25] The difference was that the British promised to offer these services on better conditions. Firstly, it advertised the introduction of important technological innovations, including new apparatus, batteries and magnets, all patented by the Highton brothers, who were the promoting

---

[22] "Electric Telegraph between Ireland and Scotland", *Morning Post*, 20 July 1852; "Electric Telegraph Company of Ireland", *London Daily News*, 29 July 1853.

[23] "The Electric Telegraph Company of Ireland is now being wound up", *Stamford Mercury*, 27 June 1856. In 1858, the company's real estate was put up for auction, but remained without buyers. "The property of the Electric Telegraph Company of Ireland", *London Daily News*, 27 August 1858.

[24] Garth Watson, *The Civils*, London, Thomas Telford, 1988, p. 251. Simpson himself emphasized the link between underground tubes and electric telegraphs when during the sitting of the parliamentary committee for the approval of the bill, he intervened as an expert in the sector: "Mr. Simpson, C. E., spoke, from the experience he had in underground works, having laid down 2.000 miles of water piping, to the inexpediency, for a variety of reasons, of laying down telegraphs upon common roads in tubes", "The Electric Telegraph Companies", *Morning Chronicle*, 28 May 1852.

[25] "From the prospectus handed to us by Mr. Challoner, sharebroker, it appears that the objects of the company are: 1. the construction of telegraphs for the government, for railway companies, and for private individuals, by contract, for a sum in gross, or for an annual payment, such telegraphs being worked and maintained by respective owners. – 2. the maintaining in efficient working condition telegraphs, whether erected by the British company or by other persons, under contracts for annual payments. – 3. the sale of licenses under the powers of the company. – 4. the construction of telegraphs to be the property of, worked, and maintained by the British Electric Telegraph Company, for the transmission of intelligence at a fixed tariff, or upon terms to be agreed upon", *Newcastle Courant*, 31 January 1851.

partners.[26] Secondly, the company declared its intention to extend its own lines along railways without telegraphs with the aim of extending the network to the North and Ireland, which in turn would be linked to English cities by means of a submarine cable.[27] In addition, unlike the Electric, all of its connections would be made via underground lines in order to avoid direct damage.[28] Thirdly, the British presented itself as an American-style telegraph company, aiming at breaking down the existing monopoly with very low flat tariffs, like those offered by the Penny Post.[29]

---

[26] "The instruments and apparatus of the *British Electric Telegraph Company* were much superior in principle, cheapness, expedition and certainty, to those used by the existing company", "The British Electric Telegraph Company", *Morning Chronicle*, 17 May 1850.

[27] "We have before us several of the company's papers in connection with this project, among others, their Prospectus, and their report presented to the first general meeting held on the 12 instant. By these we learn that, upon the old electric telegraph company having come to the determination of expending no further capital on the construction of new telegraphs throughout the United Kingdom, leaving a large portion of Great Britain and the whole of Ireland completely destitute of this new and indispensable advantage, the present company was formed and its incorporation sanctioned by the Legislature, with the express intention of providing telegraphic communication for those districts hitherto most requiring it, or absolutely in want of it; and as Ireland, especially, was left without any telegraphic communication, whether internally, or in connexion with Great Britain, the attention of the directors was turned to the establishment of a line which should afford intercommunication between the principal commercial, manufacturing and seaport towns of Ireland with those of England. The Directors have selected that part of the channel between the North east coast of Ireland and the opposite coast of Scotland as the most convenient, if not the only practicable, route for the transmarine portion of their line", "The British Electric Telegraph Company – Selection of Belfast as the Starting point for Ireland", *The Belfast News-letter*, 29 November 1850; "When the present Company was originated, the Electric Telegraph used only 2,215 out of 5,996 miles of Railways which were open and in working operation in the United Kingdom; and in Ireland, at the present moment, there does not exist a single mile of Electric Telegraph", Advertisement, *Worcestershire Chronicle*, 23 April 1851.

[28] "Powers were intended to be taken to place the wires underground, so as to be out of the reach of damage, and from the effects of atmospheric electricity – a great drawback under the present system", "The British Electric Telegraph Company", *Morning Chronicle*, 17 May 1850.

[29] "They further propose to bring the Telegraph within the reach of the humblest person, as in America, so as to have messages transmitted at not more than one-third the present charge and, as far as possible, to assimilate the system to a postal arrangement", "The British Electric Telegraph Company – Selection of Belfast as the Starting point for Ireland", *The Belfast News-letter*, 29 November 1850; "Its main object appears to be to bring up the transmission of intelligence to the American standard, which it is obvious cannot be done, without assimilation of charges to the American tariff. There cannot be a doubt that high charges, rendered unavoidable by the dead weight of a large unproductive capital have been chiefly instrumental in impeding that natural

Many of the British's claims did not really have any solid basis. First of all, though the Highton brothers' patents had been created in order not to infringe upon the Electric's,[30] they referred principally to needle and printing telegraphs not presenting any technological advance capable of revolutionizing service costs and efficiency.[31] Secondly, the company had not as yet made any formal agreements with the railways, but had only carried out some preliminary evaluations, without considering the real direct and indirect influence of the Electric on the sector. Furthermore, the British had not calculated that the fixed costs for a submarine cable and an entire underground network were far higher than those of an overhead network. And its authorized capital stock only stood at £100,000, far less than that of the Electric.[32] Finally, as emerged from some contemporary analyses, it was clear that it would be difficult to sustain a lower tariff than the Electric over a long period, given the very nature of the urban connections in Great Britain.[33]

The British was not actually capable of following its claims with deeds. In the 18 months following parliamentary authorization, it did not manage to make any of its lines operative, though continuing to announce openings in the newspapers.[34] For example, at

---

growth of the system and the great obstacles in the way of a remunerative return to the proprietors; and it is satisfactory to observe that the present company are fully alive to this important point", *Newcastle Courant*, 31 January 1851.

[30] The Highton brothers knew to perfection all the modifications of the patents possessed by the Electric, as emerges from their notes on the court cases about their presumed violation by some inventors, IET, Papers of Messers Highton 1840–1848, SC, Mss 010.

[31] Edward Highton, *The Electric Telegraph: Its History and Progress*, London, John Weale, 1852, pp. 87–101.

[32] *An Act for forming and regulating the* British Electric Telegraph Company, *and enable the said Company to work certain Letters Patent* [29 July 1850], 13–14 Victoria, c. lcccvi.

[33] "It is usual to contrast the cost of telegraphic messages with those in America; but the comparison may mislead us. The expensiveness of working telegraphs, after the necessary apparatus is formed, depends not upon distance, but upon the number of breaks or stations interrupting the course of the message, at each of which clerks must be employed and from this cause, the distances are practically greater in populous England than in America with its cities and towns few and far between. In America too the electric telegraph is useful to a larger proportion of the people and hence comes into more general requisition, not confined chiefly to parties interested in the hourly fluctuating changes of the markets. In England the telegraph outstrips other means of communication only by a few hours; in America its advantage must be measured by days or even weeks", "Electric Telegraph Companies", *Leeds Intelligencer*, 15 March 1851.

[34] "It is stated that the British Electric Telegraph Company intends to commence operations for laying down a submarine line of communication between Scotland and the

the beginning of 1852 the line between Liverpool and Manchester had been built, joining the two central offices located in the stock exchange.[35] Nevertheless, before starting the service, the British intended to complete the connection between Liverpool and London, a project that proved impossible, given that the Electric had obtained the exclusive rights for the railways surrounding London. And in fact the directors of the British had made a formal application to the board of the London & North-Western Railway Company for permission to install its lines along the latter's railways, evidently in an attempt to avoid the Electric's exclusive rights and connect Liverpool and Manchester to London via another route. But though the London & North-Western had declared it was in favour of drawing up a wayleave contract with the British, it suddenly changed its mind. For while the two companies had been preparing the documents for

North-East coast of Ireland, first forming a line from Dublin to Belfast, and thence across the Channel, which is there narrow, and so very deep as to render it but little likely that the line could be disturbed by passing vessels, or any other contact. It is said that the Lord Lieutenant views with favour these proposed operations as an excellent project for communication between the great commercial towns of Great Britain, Belfast and Dublin, and thence to the other large cities and towns of Ireland", *Morning Post*, 2 December 1850; "The British Electric Telegraph Company have taken rooms over the Royal Exchange. Their first enterprise is a great line from London by Manchester to Glasgow, thence by submarine telegraph to the Irish coast, and then through Belfast and Dublin to Cork. Other two lines are those from London to the west and the south-east", "The British Electric Telegraph Company", *Cheltenham Chronicle*, 26 December 1850; "We are happy to announce that the British Electric Telegraph Company are now engaged in preparations for laying down their line from Liverpool to Manchester. Serious obstacles yet exist in regard to the extension to London, but the use of the railway line cannot long be resisted", "British Electric Telegraph Company", *Blackburn Standard*, 11 June 1851; "We understand that offices well adapted for their operations have been taken by the company in Manchester, and they are in treaty for suitable premises in Liverpool also. It is anticipated that the line will be open for the transmission of messages between Liverpool and Manchester in from two to three months from the present time, while the directors are on the eve of concluding arrangements with several railway companies which will enable them to commence, without delay, their works from London to Liverpool. It is the expressed intention of the company not only to introduce a lower system of charges, but the adoption of instrumental improvements and inventions", "British Electric Telegraph Company", *Manchester Courier and Lancashire General Advertiser*, 5 July 1851.

[35] "The British Electric Telegraph Company have commenced laying down their system of telegraphic communication uniting Liverpool and Manchester, and the Yorkshire towns of Halifax, Bradford, Huddersfield, and Leeds, with Hartlepool, Stockton, Hull, and other north-eastern ports. Their offices in Manchester and Liverpool are close to the Exchanges in those towns, and convenient for business", "The Electric Telegraphs", *Fife Herald*, 8 January 1852.

the contract, the Electric had intervened with a convincing financial proposal which persuaded the London & North-Western to accept exclusive rights with it.[36] Deprived of its only possible telegraphic access to London, the British was again backed by the press as it started up a campaign against the exclusive rights of the Electric, with the support of numerous businessmen who sent petitions to Parliament in which they called for a new Bill for the British.[37] And the British itself was openly asking for railway companies to allow the construction of telegraph lines along their tracks,[38] requesting

---

[36] "On the 13th day of December 1850, your directors made application to the board of the London and North-Western Company for permission to use their lines of railway for this purpose, and on the 18th of January 1851 received an answer from the secretary, expressing the perfect willingness of the London and North Western Railway Company to accede to that request, and at the same time stating that they were desirous of affording to this company every facility consistent with the safety and convenience of their general arrangements. Your directors having urged the board of the London and North Western Railway Company to name the annual payment that would be satisfactory, a communication was received from the secretary, on the 17th of March 1851 setting forth in detail the terms on which free use of all their lines of railway to be conceded. These terms were accepted by your directors; but while waiting for the formal confirmation of them by the London and North Western board, they suddenly received a communication, to the effect that the London and North-Western Company had resolved not to enter into any new or permanent engagements with either Electric Telegraph Company at present, and directing all negotiations on the subject to be suspended – an announcement which virtually amounted to a repudiation of the arrangement entered into with this company, upon the faith of which being carried out, your directors had placed the fullest reliance and had made their arrangements accordingly", extract from the half-yearly report of June 1851 of the British, in "British Electric Telegraph", *Morning Chronicle*, 1 September 1851.

[37] "The extension of the system of electric telegraphs throughout the Kingdom is now regarded as a point of paramount importance, and the interest felt by commercial men in the promotion of a wholesome system of competition, with a view to preventing the entire telegraphic arrangements of the country from being directed by one single association, is sufficiently proved by the numerous petitions that have been sent up from London and the various provincial seats of commerce and industry in favour of the bill now before parliament for extending the powers of the British Electric Telegraph Company. In this sense petitions most numerously and influentially signed have been received from the London Stock Exchange, Corn Exchange, Coal Exchange, and the Ship's Towing Company; from Southwark and Finsbury; from the merchants, bankers, and ship-owners of Liverpool, Manchester, Leeds, Hull, Newcastle-on-Tyne, North and South Shields, Bradford, Halifax, Huddersfield, and Glasgow; from the members of the Stock Exchanges of Liverpool, Manchester, Glasgow, and Edinburgh; and from the Belfast Chamber of Commerce. The petitioners urge that trade and commerce have been seriously injured by the monopoly of telegraphic intelligence; and that the bill of the British Electric Telegraph Company would do away with this monopoly", *London Daily News*, 12 March 1852.

[38] *Morning Chronicle*, 13 March 1852.

authorization for an increase in capital and an extension of its limited liability. In this way a battle was started in Parliament, and the sittings of the committee appointed to examine the Bill, saw arrayed on opposite sides two major lobbies. Side by side with the British, the businessmen of the North and the press, and together with the Electric, the railway companies.[39]

The British claimed the right to lay at least one wire along the railways where the Electric had already installed its own lines. Formally, the land on which the tracks were laid was private, a state concession for the exercise of a public service. Consequently, if a company made a request to offer another public service on the same land, the Government was forced to allow it. The British had the acumen to move the attention from legal detail to a subject of great public interest: the battle against monopolies. In this way, a request that would have appeared as an evident violation of the rights of railway companies, was instead hyped as the only legal way to avoid the use of the exclusive rights with which the Electric had been eliminating competition.[40] And the hype went on, with all the businessmen and newspapers called to bear witness in favour of the British avoiding

[39] "A deputation from railway companies against the British Electric Telegraph Company had an interview with the Right Hon. J. W. Henley, yesterday, at the railways department of the Board of Trade. The deputation consisted of Capt. Tyndale and Mr. B. Oliveirs, from the Oxford, Worcester and Wolverhampton Railway Company; Mr. Stable, from the Manchester, Sheffield and Lincolnshire Railway Company; Mr. Ricardo and Mr. Dicey, from the North Staffordshire Railway Company; Mr. Venables, Mr. McGregor and Mr. Freeman, from the South Eastern Railway Company; Mr. Coates, from the Railway Clearing House and Mr. Bidder", *London Daily News*, 13 March 1852.

[40] "The petitioners set forth their impression that trade and commerce would be injured by a monopoly of telegraphic intelligence, and that the proposed bill was calculated to do away with this monopoly, which arose from the circumstance of the Electric Telegraph Company, incorporated in 1846, having entered into private arrangements with the railway companies for the exclusion of the British Telegraph Company, and all other telegraph companies, from their lines. In consequence of these agreements, the British Company was at present unable to continue their wires to London, which they were desirous of connecting with the North of England, by using the railways intervening. In the bill it was proposed to secure the railway companies from any damage, and to give them remuneration for the use of their lines, and the railway companies had obtained possession of private property by special acts of parliaments solely on the ground that it was to be used for the benefit of the public. ... The Electric Telegraph Company had, by entering into contracts at low prices with the railway companies, and by volunteering other advantages, obtained the exclusive privilege of running over the railways. This exclusion to the establishment of competing lines acted disadvantageously for the public", "Committees. The Electric Telegraph Companies", *London Daily News*, 20 May 1852.

going into the details of the Bill, but highlighting the delays, errors and extremely high cost of the telegraphic correspondence offered by the Electric. For example,

> Mr. Kershaw, stockbroker of Leeds and member of the Leeds Stock Exchange, examined, deposed that, in the opinion of the members of that body, Leeds was imperfectly served by the existing telegraphic communication. Complaints were constantly occurring, on the score of inaccuracy and delays arising from pressure of despatches.
>
> Mr. J. Greig, proprietor and editor of the *Scottish Press* newspaper, had frequently had his attention drawn by the commercial and municipal bodies of Edinburgh to the present state of their telegraphic communication, and knew that their feeling was one of dissatisfaction regarding it.
>
> Mr. T. Read, of the Stock Exchange, Glasgow, spoke to the way in which the telegraph service was performed for that community. It was not satisfactory.
>
> Mr. A. Ashton, stockbroker of Liverpool, and member of the Liverpool Stock Exchange, deposed to the dissatisfaction felt by that body with respect to the present system of telegraphic service between Liverpool and London.
>
> Mr. G. V. Robinson, London corresponding agent to Messrs. Willmer and Smith, of Liverpool, and to the American and Continental newspapers, had great reason to complain of the manner in which, since 1848, the business was conducted by the Electric Telegraph Company, and as to errors, delays and excessive charges, knew that Messrs. Willmer and Smith had suffered loss thereby.
>
> Mr. Cowan, M. P. for Edinburgh, spoke to the complaints that existed among his constituents with reference to the excessive charges, delays, and inaccuracies connected with the electric telegraph.
>
> Mr. J. S. Challoner, stock-broker of Newcastle, deposed to the dissatisfaction existing there, and adduced instances of inaccuracy and delay.[41]

On their part, the Electric and the railway companies defended themselves by trying to emphasize the illegality of the British's request and how it could damage their business activity. The Electric started by pointing out that as first mover it had all the rights to establish exclusive contracts with railway companies. Ricardo then recalled that contracts of this kind existed for many different types of activity and their validity had never been doubted. He gave the example of a railway station, that no one expected it to concede

---

[41] "Parliamentary Committees. The Electric Telegraph Companies", *London Daily News*, 24 May 1852.

spaces to other applicants beyond the refreshment vendor offering the best service.[42] The railway companies found it unacceptable for other private parties to dispose of their structures without their consent, just by paying rent.[43] To defend themselves they gave legal reasons and also expressed their fears about safety, in that new telegraph installations erected without previous agreement with the railways could cause numerous accidents.[44]

It was, however, more difficult for the Electric to defend its monopolistic stand with public opinion, especially because of the current battles in favour of free trade (the most famous being the anti-Corn Law Campaign). Two lines of defence, starting from opposing economic assumptions, were used to defend Electric's grasp over the market.

On one side, Ricardo, who as a liberal and defender of free trade, claimed the role of first mover for the Electric. To reach this position

---

[42] "The company's arrangements with the railway companies for the exclusive use of the line were fair and reasonable; the stipulations being identical with the ordinary covenants between shopkeepers or taverns in the same street or on the same estate. Could not see why any bookseller, newsvendor, or refreshment vendor should not insist upon compulsory power to enter any railway station, as for a telegraph company to do the same", "The Electric Telegraph Companies", *Morning Chronicle*, 2 June 1852.

[43] "Mr. Alexander, Q. C., in a lengthy address, opened the case in opposition to the British Telegraph Company's bill on behalf of twenty-seven railway companies, on the ground that the bill sought to invest the promoters of it with power to enter compulsorily upon any railway or canal in order to lay down telegraphs and apparatus for their own private benefit. The bill also proposed to empower the promoters to remove, either permanently or temporarily, as occasion or convenience might require, in respect of their works, all obstructions and impediments that might exist on all railways or canals in the United Kingdom. The railway companies he represented opposed the delegation of any such compulsory powers to any undertaking, as novel and altogether anomalous", "The Electric Telegraph Companies. Case for the Railway Companies", *Morning Chronicle*, 1 June 1852.

[44] "Mr. G. R. M'Clean, Civil Engineer, who stated that he hold a lease of the South Staffordshire Railway for twenty years, and that he had at his own expense constructed a telegraph on that line. Great inconvenience was felt, and much care required, in adjusting and repairing the telegraph at present on the railway, and an additional telegraph would double the inconvenience and the nuisance, independent of the danger that would accrue to the safe working of the railway if compulsorily entered upon by rival parties", Ibid.; "Mr. Hawkshaw, Civil Engineer, held two shares in the Electric Telegraph Company. Was of opinion that would be required for a second telegraph, more particularly as, from the constant development of the traffic, they now found their present space inadequate. Believed that to confer the compulsory powers sought for would be to strike at the root of railway safety, and he would not be responsible for the security of a railway under the circumstances. Looked on the proposed powers as an invasion of the right of railways", "The Electric Telegraph Companies", *Morning Chronicle*, 4 June 1852.

the company had had to bear heavy initial costs, especially for the purchase of patents and wayleave rights, which later rivals had not had to face. Ricardo felt, therefore, that the Electric was totally justified in setting up barriers against entry, like the exclusive rights accorded with the railways in order to hinder the entry of late comers. In other words, Ricardo was explaining he did not want to exclude the British and other companies from the market and was quite ready for them to open telegraph communications along railways not covered by the Electric and connect places where the Electric's lines already existed, by however installing lines along canals and highways. In the latter case, obviously, the law called for underground lines, which meant higher fixed costs, but Ricardo felt this was just compensation for late comers avoiding start-up expenses.[45] To drive its point home, the Electric asked some of the top electric engineers of the time to explain to the committee that underground lines were as efficient as overhead lines, if not more so.[46]

Differently, other defenders of the Electric, including Stephenson, argued that two telegraph lines along the same route were

---

[45] This was what Ricardo argued in front of the committee: "Believed that competition of telegraphs on railways would resolve itself merely into a question of charge, that would result in the ruin of the weaker party; but that competition of a legitimate character might be maintained by a system of telegraphs on common roads, which would accommodate places with telegraph not served by railways. Thought the construction of telegraphs on common roads feasible, and that it would not be more costly than the suspension system upon railways. ... The contract the Electric Telegraph Company had entered into with the railways was an exclusive contract, just as every other contract for permanent way and other purposes of railways was an exclusive contract. Believed, from a calculation he had made, that it would be cheaper to carry a telegraph by common road than by railway, looking at the cost of rent, way-leave, and repairs", "The Electric Telegraph Companies", 2 June 1852.

[46] "Mr. Edwin Clark, C.E., engineer-in-chief to the Electric Telegraph Company, stated that he had, within the last twelve months, put up 1,500 miles of telegraph, and gave it as his opinion that the underground system of telegraphs was perfectly practicable. The best and most perfect insulation for the wires underground was gutta percha, with which they were supplied by the Gutta Percha Company, whose patent had also been successfully applied for the submarine telegraphs. ... Mr. Gilpin, contractor for the Irish Channel submarine telegraph, now being carried out between Portpatrick and Donaghadee, described the operations in connection with that undertaking. About sixty miles of the eighty-five of trench for the underground telegraph on the Scottish coast was finished. ... Thought the underground was better than above-ground system", "The Electric Telegraph companies", *Morning Chronicle*, 3 June 1852; "Mr. C. W. Siemens, C. E., spoke to the system of sub-humous telegraphs in Prussia. The first cost of the underground system, of which there were 30 miles in Berlin, would be dearer than the suspensory system, but quite as efficient", "The Electric Telegraph Companies", *Morning Chronicle*, 4 June 1852.

economically useless, as would be two railway tracks. Installing two rival telegraph lines on the same route would cause a notable disservice to the public, which would have to do without the opening of new lines because of the fixed and variable costs entailed for duplicating the same line.[47]

This position implied backing the so-called natural monopolies: i.e., where supplying a service involved building costly infrastructures, it was more expedient to avoid competition for the sake of the users.[48] The positions of Ricardo and Stephenson were notably divergent, though both were aiming at defending the same company. Once the bitter rivalry with the British had calmed down after 1855, these two opposing economic visions emerged with greater force and were among the main causes of the re-shuffle of the Electric's board of directors.[49]

---

[47] Stephenson's opinion as expressed in front of the committee as follows: "competition as applied to telegraphs, or, in other words, competition between lightning and lightning, was an absurdity. It had been proved to be an absurdity as regarded railways, for competition always closed in a coalition; railways could never be worked so well as when centralized under one establishment, and the result was, as it would be with telegraphs, that the public would derive the greater benefit. In telegraphs almost nine-tenths of the expense arose from stations and clerks, and it would be for the interest of the public, as regarded cost, instead of having double establishments, to have them under one establishment. While of opinion that competition in telegraphs was a perfect absurdity, thought the telegraph companies ought to be compelled periodically to come before Parliament and have their rates revised when dividing more than a given percentage, and that that would be the very best mode, both in telegraphs and other undertakings, for preventing the reckless waste of capital by millions", "The Electric Telegraph Companies. Case for the Railway Companies", *Morning Chronicle*, 1 June 1852. For a similar interpretation: "Mr. Locke, M. P., was the next witness examined. ... The public interest was identified with having both the railway and telegraph service of this country done in the best possible way, but did not believe that the interest would be promoted by doubling the number of lines of wire, or the persons employed; for if the companies, as was frequently the case, combined afterwards, the public would have to pay for the outlay of two capitals and the cost of two manipulations, instead of one. Would make one telegraph do all the work, just as he would one railway; and the public, if they only saw that, as the result of competition, they had to pay in the end double, would, he believed, think so too", Ibid.

[48] It was postulated indeed by the very influential John Stuart Mill that natural monopolies are situations in which the running of public services is more efficient if carried out in the absence of competition. He had in mind public services like water and gas, which call for the construction of big infrastructures, and consequently, high fixed costs. As natural monopolies, they allow the economies of scale to be exploited and avoid the duplication of infrastructures on the same territory. John Stuart Mill, *Principles of Political Economy*, Vol. II, Boston, Charles C. Little & James Brown, 1848, pp. 537–557.

[49] Roger Neil Barton, "New Media. The Birth of Telegraphic News in Britain 1847–68", *Media History*, 16/4, 2010, pp. 379–406.

As was foreseeable, thanks to the backing of public opinion the British got the better of the situation. The committee declared its requests were legitimate, but excluded from the Bill its request for limited liability, which needed economic guarantees it had not presented.[50] Parliament then ratified the committee's judgement and passed the Bill, but as a victory it was ephemeral.[51] The Lords recognized a formal error in the presentation of the Bill and invalidated the whole procedure.[52] As pointed out by a consultant during a committee session, in asking for compulsory powers the British should have deposited three quarters of the foreseen expense.[53] In spite of the British's vigorous protests and the usual press attack on the Electric, the Bill was never sent to the Commons.[54] The British had to await its merger with the European in 1854 in order to include London in its network.[55] Meanwhile, it carried out a full restyling

---

[50] "Parliamentary Committees, The Electric Telegraph Companies", *London Daily News*, 10 June 1852.

[51] *Morning Chronicle*, 18 June 1852.

[52] "Thrown out upon the ground that 'Standing Order n. 176 of the House of Lords' was applicable, and had not been complied with ought not be dispensed with" [Advertisement.] "British Electric Telegraph Company Bill", *London Standard*, 25 June 1852.

[53] "Mr. Coates, of the firm of Dyson, Parkes, Coates and Co., parliamentary agents, deposed, from long experience, that it was the general rule of Parliament, established by the standing orders of both houses, not to grant compulsory powers over other men's lands without a plan defining precisely the property to be taken, an estimate of the cost of the undertaking, a subscription contract for three-fourths of the amount of the estimate, a deposit before the introduction of the bill of one tenth in the case of railways and one-twentieth in other cases of the amount subscribed, with lists showing the opinions of the persons whose property is to be taken. By 8Vic., c., 18, a company was prohibited from exercising its compulsory powers until the whole of its capital was subscribed, or from entering upon lands compulsorily without payment of the price. The Bill of the British Telegraph contained no such provisions. In no case were compulsory powers granted by Parliament without the tools or rates being defined in the bill, and he was of opinion that there was no precedent whatever for such a bill as that proposed by the British Company", "The Electric Telegraph Companies", *Morning Chronicle*, 2 June 1852.

[54] Ibid.

[55] "The attention of the directors was next anxiously turned to the completion of the communication to London, in the speediest and most effectual manner. At this juncture the European and American Electric Printing Telegraph Company had just completed the laying down of their wires from Dover to London, and thence to Liverpool and Manchester and a correspondence having been opened with that company they were found willing to unite their lines with those of the British Company, thus in connexion with their own northern lines, supplying to the latter, without the delay of construction, the communication from Liverpool, Manchester, Glasgow, Hull, Newcastle and intermediate stations to London, and at the same time placing those districts in direct communication with the continent of Europe", "British Telegraph Company", *London Daily News*, 4 September 1854.

to appear more attractive to possible shareholders and investors. Its first step was to get a new Bill passed authorizing it to increase its capital.[56] It then wound up the company and refounded it under the name British Telegraph Company with the dual aim of setting itself clearly apart from the Electric and furnishing the public with an image of a new company.[57] The break with the past was also emphasized by the way the British gave a new numeration to its half-yearly meetings, so distancing from an inoperative past.[58] All these changes were successful, for the number of shareholders rose appreciably, reaching 169 by September 1854. It was a broad-based holding, with most of the participants coming from Glasgow, another consistent group from Manchester, including John Pender, a main player in submarine cables from the 1870s onwards, while the others were distributed among the cities in the British's network, like Hull and Bradford. Significantly, only one shareholder came from London. The most substantial quota was in the hands of three: Samuel Laing, then chairman and managing director of the London, Brighton & South Coast Railway, chairman of the Crystal Palace Company and Liberal member of Parliament, Ernest Christian Louis Bunsen, a writer of German origin and Lord de Mauley, chairman of both the Submarine and European.[59] All three were on the boards of the Submarine and European, so that the block of shares may well have been inherited from the British takeover of the European, which had taken place a few months earlier.

Probably because of the new entries, data referring to the correspondence service emerged at once, during the first shareholder meeting in 1855,[60] witnessing in an explicit way that the company was fully operational. As the great rival Ricardo was quick to point out, in spite of all the claims and massive press advertising campaigns, in its first four years of existence the British had built very few miles of telegraph lines with a network which never got further than outer London. Furthermore, the service it offered was quantitatively

[56] *An Act for repealing and amending an Act passed in the Thirteenth and Fourteenth Years of the Reign of Her present Majesty, called "The British Electric Telegraph Company's Act"* [4 August 1853], 16–17 Victoria, c. clix.
[57] BT, *British and Irish Telegraph Company*, TGL 1/1/1, *Deed of Settlement*, 1854.
[58] "British Electric Telegraph Company", *Paisley Herald and Renfrewshire Advertiser*, 2 September 1854.
[59] BT, *British and Irish Telegraph Company*, TGL 1/1/1, *Shareholder List*, 1854.
[60] *Morning Chronicle*, 3 March 1855.

very limited, unlike some rival companies which had steadily grown in size and in silence.[61]

3.3   The Magnetic: low profile but high technology

After having announced its parliamentary authorization as well as those of two other companies in summer 1851,[62] the press almost lost interest in it, except to publish every now and then the construction of new lines.[63] Unlike the British, the Magnetic moved pragmatically in its first year and dedicated itself to the installation of a telegraph network, avoiding both optimistic claims and a direct opposition to the Electric. Proceeding in the same way, the Magnetic began its telegraph service in the second half of 1852, between Liverpool, the company's operative centre, and Manchester. At the same time it

---

[61] In front of the committee, Ricardo gave his opinion: "The British Company, during the first two years of their existence, had laid down none, though the Magnetic-Electric Telegraph Company, that started a year after the British Company, had already done a great deal. The British Telegraph Company had made large professions, but no performance. They proposed communication with the Continent and Ireland two years ago, but other companies had carried it out. They talked about domestic telegraphs from the dining room to the kitchen, but he did not understand that they had made any progress in identifying the electric telegraph with the hearts and homes of England", "The Electric Telegraph Companies", *Morning Chronicle*, 2 June 1852.

[62] "There are five applications for new electric telegraph companies – viz: The United Kingdom Electric Telegraph Company (Allan's patent), the European and American Printing Telegraph Company (Jacob Brett), the Submarine Telegraph Company between Great Britain and Ireland (J. Brett), the Submarine Telegraph Company between England and France, and the Magnetic Electric Telegraph Company. The Electric Telegraph Company also apply for amendments of their present acts", *Parliamentary Applications for the Ensuing Session*, "London Daily News", 2 January 1851. Only the Magnetic, the American and the United obtained permission on this occasion. "Parliamentary Committees", *London Standard*, 3 June 1851.

[63] "The Magnetic Telegraph Company's Act (Henley's patent) received the Royal assent on the 1st. The offices of the company are to be in Liverpool, and it is said that extensive works, under the superintendence of Fox, Henderson, and Co. are to be commenced immediately. The first line to be laid down is the Lancashire and Yorkshire", *Bradford Observer*, 2 August 1851; "Upwards of 200 miles of insulated wire have already been laid down by the Magnetic Telegraph Company upon a portion of the Lancashire and Yorkshire Railway; and the new line of Telegraph between Manchester and Liverpool, belonging to his company, will be opened in a week or two. The Magnetic Telegraph appears to have fulfilled all the favourable anticipations that have been formed respecting it, wherever it has been employed", "The Magnetic Telegraph", *Durham County Advertiser*, 21 November 1851; "A charted has been granted to the English and Irish Magnetic Telegraph Company. By this line the Home Office, in London, will be connected with the Castle, in Dublin", *Yorkshire Gazette*, 12 June 1852.

extended its network as far as London, succeeding where the British had failed, to activate a service with the capital already in the first half of 1853.[64]

On paper the Magnetic's strategies for entering the market were very similar to those of the British. It wanted to construct lines where the Electric had no infrastructures or not enough to answer the needs of the industrialists and local businessmen. So starting in Lancashire and Yorkshire, it then linked up with Scotland and Ireland, where there were no rivals in the market.[65] The Magnetic based its development on an extremely innovative technology, both in the type of telegraph adopted and the way of constructing the lines.

Henley-Foster's electromagnetic telegraph, whose rights of usufruct were held by Magnetic, was very different from those adopted by other companies. While the Wheatstone-Cooke and Highton Brothers needle telegraphs were based on the magnetic effect produced by electricity, the Henley-Foster model deployed the electric current generated by a magnetic field. Two magnets were mounted inside the telegraph and when the telegraph operator pressed the transmitter button the magnets turned over, so that their poles went

---

[64] "It is with pleasure we see that the English and Irish Magnetic Telegraph Company are coming before the public, already having quietly constructed, or made arrangements for the completion of their extensive system of telegraphs, which will at last, we think, afford what we have long needed – a sufficient and certain means of communication. Their main lines being laid underground will obviate that uncertainty which has hitherto been experienced in the working of the overground system in wet weather. The London, Manchester, and Liverpool line of this Company will be of the most approved construction; and is estimated, we think fairly, to be able to meet the whole of the requirements of the public between these the three largest towns of England. A large portion of the lines of the Company in the United Kingdom is already completed", "English and Irish Magnetic Telegraph Company", *Leamington Spa Courier*, 1 January 1853. Note how the journalist uses the adverb "quietly", to describe the low profile adopted by the Magnetic when constructing its lines.

[65] "The Belfast Whig contains the following statement of the operations of the Magnetic Telegraph Company, extending through all parts of the United Kingdom: 'We have it on authority that a very brief period will elapse ere Belfast be placed in direct communication with the three capitals of these kingdoms, as well as with the capitals of the Continent. Measures to this effect are at present in progress, and will not, we are satisfied, meet with impediment or procrastination. The scheme of telegraph actually agreed to by the company takes in Dover, London, Birmingham, Wolverhampton, Manchester, Bolton, Wigan, Liverpool, Preston, Carlisle, Edinburgh, Glasgow, and Greenock, in Great Britain; and Donaghadee, Belfast, Dublin, Galway, Limerick, Tipperary, Waterford, Cork, and the intermediate towns in Ireland'", "Great Extension of the Magnetic Telegraph", *Morning Chronicle*, 22 July 1852.

into opposition and generated an electric current. Transmitted along the wire, it reached the receiver apparatus and moved the needles right or left, with the different composition of the oscillations generating a communication code, as with all needle telegraphs.[66] The big difference was that unlike all the other telegraphs of the period, the Henley-Foster model generated its own electricity and worked without batteries.[67] Given that it was not much more expensive than the Electric's needle telegraphs and needed very little maintenance in its first six months of functioning,[68] it promised a notable reduction in costs.[69]

The Magnetic had also planned to install underground telegraph lines, which were held to be more resistant to weather and wear and tear than the traditional cheaper overhead lines adopted by the Electric. The aim was to be able to operate in extreme conditions

---

[66] "When the levers are pressed down, the electro-magnets are reversed in the relation of their poles to those of the permanent magnets, and momentary currents are transmitted on the conducting wires, and when the levers are observed to rise to their former position, momentary currents are again transmitted, but in a contrary direction. The currents thus transmitted on the line-wires are received at the station to which the dispatch is transmitted upon the coils of electro-magnets, which are placed under the desk upon which the indicating needles are placed, and they impart temporary magnetism to these. These electro-magnets act upon a small permanent magnet suspended under the desk, on the axis of the indicating needle, and parallel to it. They deflect this needle on the one side or the other, at the moment they receive the magnetism from the current, and their deflection is continued by the effect of the induced magnetism produced by a permanent magnet on the electro-magnet". Dionysus Lardner, *The Electric Telegraph Popularised*, London, Walton and Maberly, 1855, pp. 58–59.

[67] "The Magnetic Telegraph Company, retaining the needle indicators generally used in England, have rejected the galvanic battery, and substituted the magneto-electric for the voltaic current", Ibid., p. 58.

[68] "No cost is incurred in maintenance of the Telegraph Instruments, have such occasional mechanical repair at May, through accidental damage become requisite", BT, *English and Irish Magnetic Telegraph Company*, TGN 1/1/1, *Memoranda Relating to General Meeting*, 4 January 1853.

[69] "I'm strongly in favour of the use of magneto-electricity. Its economy is undoubtedly the most prominent feature. A pair of magnets, costing at Sheffield 30s. and perhaps 40s. to 45s. (according to the finish bestowed on the instrument), by the time they are fixed and ready for use, will send a strong current on a well insulated suspended line for above 200 miles, and on underground wire above 100. (I have had signals, but only weak ones, through 250 miles of underground wires with the class of instruments I am speaking of), while the six twelve-cell through battery used in this country, which would be necessary to perform the same work, would cost 7L. 10s., besides the constant expense of renewal", Charles Bright, answer n. 77, Appendix IV, Edward B. Bright and Charles Bright, *The Life Story of the Late Sir Charles Tilston Bright*, Westminster, Archibald Constable and Co., n.d., p. 440.

and guarantee great revenues.[70] In addition, now that lines could be installed along both railroads and highways, the Magnetic was not involved in the legal problems which had drawn the British into its extenuating parliamentary battle.

The Magnetic's first, provisory board, held before parliamentary authorization had been obtained, brought together several pioneers, like W.T. Henley, the inventor, the German C.W. Siemens and Hatcher, the former secretary and chief engineer of the Electric. Thanks to them, the first lines between Liverpool, Bolton and Manchester were built. Local businessmen were so interested that in 1852 a Royal Charter was obtained which allowed an increase in capital, limited liability and the renaming of the company as the English & Irish Magnetic Telegraph Company.[71] At the same time, the revenue of the correspondence service, which started operations in the spring-summer, went on growing for the rest of the year,[72] followed by a progressive increase in the number of shareholders between 17 August and 19 November.[73] At the moment of the convocation of

---

[70] "During the four winter months of November, December, January and February 1853–54, distances of 300 miles of underground wire, without any break of circuit, have been in constant operation under the Magnetic Telegraphic Company, and notwithstanding an unusual prevalence of unfavourable weather, with frequent and continued snow-storms, no stoppage whatever has taken place", Lardner, *The Electric*, p. 149.

[71] "The first general meeting of this company, which has been incorporated by royal charter, took place at their offices, 1, North John-Street, Liverpool, a few days since, and was attended by a very numerous and influential body of shareholders.... The royal charter of the company lay on the board table, and was an object of interest to many, who for months past had seen all attempts to bring out telegraph companies, headed by a promise of royal charter and limited liability. The document was the more satisfactory to the shareholders, from the fact of its being duly endorsed and certified by the proper authorities, that all its required conditions had been fulfilled", "English and Irish Magnetic Telegraph Company", *London Daily News*, 19 January 1853. The constitutive document of the company with its new features in BT, *English and Irish Magnetic Telegraph Company*, TGN 1/1/1, *Deed of Settlement of the English and Irish Magnetic Telegraph Company.*

[72] "*Satisfactory revenue* The accounts show an amount of profit for the past half year, upon the cost of the wires in use, equal to 83/5 per cent for annum. *Steadily progressing increase in receipts*. Averages struck every six weeks show that the Traffic receipts of this Line have, without exception, steadily and rapidly during the three quarters of a year that it has been open to the public – the receipts at the termination of the half year being, in averages of six weeks, double those of the commencement", BT, *English and Irish Magnetic Telegraph Company*, TGN 1/1/1, *Memoranda Relating to General Meeting*, 4 January 1853.

[73] Of the 53 partners at the end of November, 11 had bought the shares in August, another 11 in September, 18 in October and 13 by mid-November, which goes to

the first meeting in 1853, the capital was considerably split up, as there was no majority shareholder, most holding from about 50 to 100 shares. And the maximum of individual shares (159) was held by three people: Edward Cropper, director and shareholder of numerous railway companies including the London & North-Western,[74] Charles Fox, civil and railway engineer involved in the London & Birmingham and the construction of the Crystal Palace,[75] and George Wythes, a railway contractor from Reigate. Almost half the shareholders, 20 out of 53, were from Liverpool and thereabouts, while almost all of them belonged to the industrial/commercial class, 15 described themselves as merchants, 8 as brokers, 3 goldsmiths, 2 bankers and 2 as contractors.[76]

Joseph C. Ewart, a Liverpool industrialist, director of the North Union Railway, and member of Parliament since 1855,[77] was appointed chairman, flanked by Cropper as deputy.[78]

Although the numerous new shareholders only held limited quotas, they guaranteed a strong capitalization, indispensable in a moment in which the company was building the lines in Ireland, laying a cable over to Scotland and finishing the underground link with London.[79] Guiding the company in this delicate period of technical

---

show an increasing interest in the Magnetic. Data calculated from the shareholder list attached to BT, *English and Irish Magnetic Telegraph Company*, TGN 1/1/1, *Deed of Settlement of the English and Irish Magnetic Telegraph Company*.

[74] Kieve, *The Electric Telegraph*, p. 54.

[75] Sir Charles Fox (1810–1874), Robert Thorne, *Oxford Dictionary of National Biography*, 2004, accessed 16 November 2014.

[76] The data comes from the list of shareholders attached to BT, *English and Irish Magnetic Telegraph Company*, TGN 1/1/1, *Deed of Settlement of the English and Irish Magnetic Telegraph Company*.

[77] Kieve, *The Electric Telegraph*, p. 54.

[78] "English and Irish Magnetic Telegraph Company", *London Daily News*, 19 January 1853.

[79] "*London Line*. With reference to the mode in which this Line is to be constructed, the underground system has been selected, generally, for those portions of the main route in England and Scotland uncontracted for prior to the commencement of the half year, the experience derived from the successful working of the Manchester and Liverpool Line (which has never been out of order save from accidental damage) showing that this plan offers the best, surest, and most permanent investment for the shareholders, and that any difference in cost is quite counterbalanced by the greater efficiency of the underground and compared with the above ground system. The Manchester and Liverpool Line having, in no instance suffered from defective insulation or those atmospherical effects which the suspended wires of the Electric Telegraph Company have during the past half year, been constantly subject to", BT, *English and Irish Magnetic Telegraph Company*, TGN 1/1/1, *Memoranda Relating to General Meeting*, 4 January 1853.

development and modification of the ownership was Edward Bright, in the role of secretary.[80] A little over 20 at this time, Edward was a distant relation of Cooke's and had worked between 1848 and 1851 for the Electric, together with his younger brother, Charles.[81]

Charles Bright was taken on as chief engineer in 1852, thanks to the intervention of the shareholders and directors Robert Crosbie and Henry Harrison.[82] The previous year he had invented for the British a new method of installing underground lines. Instead of threading the gutta-percha covered telegraph cables inside steel tubes designed for gas distribution, he placed them on the lower part of steel tubing cut in half and then closed it. In this way, the rubber coating was protected from being damaged while being inserted in the tubing, which so often happened with the Electric's traditional method.[83] Bright used his new method for the first time when wiring the city of Manchester for the British.

---

[80] BT, *English and Irish Magnetic Telegraph Company*, TGN 1/1/1, *Deed of Settlement of the English and Irish Magnetic Telegraph Company*.

[81] "Charles Bright and his brother Edward...joined, when respectively fifteen and sixteen years old, the Electric Telegraph Company, under the auspices of Mr. (after Sir William) Fothergill Cooke, who was a connection by marriage", Edward and Charles Bright, *The Life Story*, pp. 35–36.

[82] "In 1852, [Charles Bright], when scarcely twenty years of age, was asked by the Board, at the instigation of Mr. Crosbie and Mr. Harrison of the Managing Committee, to become their engineer-in-chief, which post he accepted, resigning his position on the British", Ibid., p. 49.

[83] "Street wires used to be drawn through solid gas piping of about 3 inches diameter, the pipes being laid first, and the insulated wires drawn afterwards. In doing this the insulating material was frequently injured; sometimes the wires were broken inside the gutta-percha or other insulating material by the force necessary to pull them trough, and occasionally they were drawn so tight that on the slight settlement of the ground, usual after the line had been laid a short time, some of the wires broke inside the insulating material, occasioning great difficulty and expense in detecting the fault. The great proportion of the faults, however, were only abrasions of insulating material; and though at the time the wires passed with all appearance of perfection through the ordeal of testing, and the streets were closed, and the pavements reinstated, before long the defects became so manifest as to interfere with the working of the apparatus, and the streets had to be re-opened, and the wires tested through length by length for the fault. The wires required jointing at every other drawing point, and these points frequently proved defective, particularly in the old varnished cotton method of insulation and others prior to the use of gutta-percha. In the beginning of 1852, having a considerable lengths of street work to lay, I gave a good deal of attention to the subject, and determined on having the pipes cast longitudinally in two pieces, so that the wires could be *laid in* the under lengths, and the upper lengths then attached, instead of drawing or threading them though solid pipe", Charles Bright, answer no. 26, Appendix IV, Edward and Charles Bright, *The Life Story*, pp. 434–435.

Very much aware that he would have to work quickly and fault-free so as not to slow down the intense mercantile and industrial activity of Manchester, Charles set out a detailed work plan which divided the installation process into four phases, each carried out by different squad. The first one dug up the road and laid the first half of the steel tube in a little trough. Then the second squad placed the cable in the open half of the steel tube and added a layer of tar to increase insulation. The task of the third squad was to close the steel tube, while the last one filled in the road and smoothed over the surface. The whole operation was carried out in a night, and the young Charles received his first praise from the press, which would always place him, for good or evil, at the centre of all the events concerning submarine cables.[84]

Although the documents do not give the explicit reasons for his move from the British to the Magnetic, at least three factors may well have weighed heavily on his decision: a personal desire to oppose the Electric's monopoly by building an alternative

---

[84] "Young Bright was first brought into public notice by a remarkable feat, namely, the laying of the Manchester underground wires in 1851. It was essential that the traffic of so busy a city should be interrupted as little as possible. Charles Bright did not interrupt the traffic at all. In one night he had the streets up, laid the wires, and had laid the pavements down again before the inhabitants were out of their beds in the morning. He was then but nineteen, and received great credit in the public journals, notably in *The Times*, which made this piece of work the subject of a leading article. 'The following arrangements for the night's work go to show the prescience and energy characteristic of him. Iron pipes were cast in halves, longitudinally, with side tongues and clips. The gutta-percha covered wires were wrapped with tarred yarn into a rope, and wound on broad drums with huge flanges. A large number of navvies were engaged, with competent foremen. To each gang was assigned a given length of street, along which the flagstones were to be lifted, the trench opened to the requisite depth, and the under-halves of the pipes previously laid down. A further gang followed for applying, linking, and tightening the upper-halves of the pipes, while yet another set of men filled up the trench and replaced the flags'. This operation though easily described, required at this early stage of telegraphy a great deal of consideration, coupled with very active and determined control throughout the short night. Charles Bright subsequently carried out the same system in London, Liverpool and other large towns", Edward and Charles Bright, *The Life Story*, pp. 46–48.

[85] "It will not be unpleasant to me – however uncomfortable generally and disagreeable in retail – for as you know, my aim for some time has been to weave a web of wire in opposition to the monopoly, and as I cannot do it for ourselves, I am well content to do it for others. Having no stake or responsibility in it I felt more comfortable perhaps than I should have had we succeeded in establishing a Company, which would have been a case of either make or mar", "Letter of Charles Bright to his fiancée on 5 September 1851", cited in Ibid., p. 45.

network;[85] the long period of the British's inactivity during the legal battles; the changed ownership of the Magnetic and its recapitalization. And in his powerful managerial position in the Magnetic, Edward Bright must have influenced his brother's choice. The two obviously had a harmonious relationship, and when rival companies took them on in 1851, they had still kept in contact and spent their Sundays regularly either at Charles' house in Manchester or Edward's in Liverpool.[86] Over the years they worked together on numerous patents for implementing telegraph lines and apparatus and gave the Magnetic an edge over the others from 1852 onwards.[87] It was no coincidence that Charles's method for laying underground lines was used for the construction of all the Magnetic's main lines, including the important link with London.[88] Moreover, the only defect of the underground system was the possibility of an electric conductor breakage, which was put right by the introduction of the Bright Brothers' galvanometer.[89]

---

[86] "Thus the two brothers became engaged in advancing the early stages of two competing concerns – a curious and novel position for them. Charles's headquarters were in Manchester, whilst Edward was stationed in Liverpool. As a rule, however, each passed alternate Sundays with the other", Ibid., p. 40.

[87] "It was in this year [1852] that the brothers took out their famous patent ... It contained twenty-four distinct inventions connected with telegraphs", Ibid., p. 49.

[88] "The method of laying the wires in the streets adopted by this company is a little different. In this case iron pipes are laid, but they are split longitudinally. The under halves are laid down in the trench, and the gutta-percha covered wires being deposited, the upper halves of the pipes are laid on and secured in their places by means of screws through flanges left outside for the purpose. To deposit the rope of gutta-percha-covered wires in the trough it is first coiled upon a large drum, which being rolled along slowly and uniformly over the trench, the rope of wires is played off easily and evenly into its bed. So well has this method of laying the wires succeeded that in Liverpool the entire distance along the streets from Tithe Barn Railway station to the Telegraph Company's offices in Exchange Street, East, was laid in eleven hours; and in Manchester the line of streets from the Salford Railway station to Ducie Street, Exchange, was laid in twenty-two hours. This was the entire time occupied in opening the trenches, laying down the telegraph wires, refilling the trenches and relaying the pavement", Lardner, *Electric Telegraph*, pp. 146–147.

[89] "One of the objections against the underground system of conducting wires, was, that while they offered no certain guarantee against the accidental occurrence of faulty points where their insulation might be rendered imperfect, and where, therefore, the current would escape to the Earth, they rendered the detection of such faulty points extremely difficult. ... remedy for this serious inconvenience, and a ready and certain method of ascertaining the exact place of such points of fault without leaving the chief, or other station at which the agent may happen to be, has been invented and patented by the Messrs. Bright of the Magnetic Telegraph Company. Instruments called Galvanometers...are constructed, by which the relative intensity of electric currents is measured by their effect in deflecting a magnetic needle from its position of rest", Ibid., p. 147.

Between January 1853 and January 1855, the Magnetic's network leapt ahead. In May 1853, it stole the march on its rivals by laying a submarine cable built by R.S. Newall and Co. between Donaghadee in Ireland and Portpatrick in Scotland.[90] A few weeks later, the cable started operating officially, although the links to Irish and English cities had not been completed. At the same time agreements were made with many Irish railway companies to install overhead lines on the model of the Electric.[91] In the second half of the year the network was extended all over the country, linking in particular the cities/towns of Dublin, Belfast, Cork, Queenstown, Galway, Drogheda, Dundalk, Athlone, Ballinasloe, Portadown, Maryborough, Thurles, Newry, Mullingar and Tipperary.[92] The initial Lancashire-Yorkshire network had already been connected to Portpatrick, via a long underground line from Preston through Lancaster, Carlisle, Dumfries and Stranraer. At Carlisle the underground cable divided into two, with the second part going in the direction of Edinburgh, Glasgow, Greenock and other Scottish cities.[93]

---

[90] "The curve experiments having been completed, and the integrity of the insulation of the wires having been proved beyond a doubt, the first formal message was transmitted to Ireland, on Monday morning, as follows: 'Mora Bay, Portpatrick, Scotland, 23rd May, 1853. The directors of the English and Irish Magnetic Telegraph Company beg to acquaint his Excellency the Lord Lieutenant, that they, this morning, successfully effected communication between the shores of Great Britain and Ireland, by means of a submarine cable from Portpatrick to Donaghadee'", "The Magnetic Telegraph. Laying down of the cable across the North Channel", *Morning Post*, 26 May 1853.

[91] "Ireland is likely soon to be overspread with an immense network of telegraphs. At present, as I perceive from the files of *The Whig*, the Magneto-Electric Company have entered into contracts for the Dublin and Drogheda Railway, 32 miles; the Navan Branch, 19 ½ miles; the Kells Branch, 9 ¾ miles; the Dublin and Belfast Junction, 56 miles; the Ulster Railway, 25 miles, Portadown to Belfast, and Portadown to Armagh, 11 miles; the County Down Railways to Newtownards and Holywood; the Great Southern and Western; the Kildare and Carlow; the Killarney Branch", Ibid.

[92] "The cities and towns brought into telegraphic *rapport* by this extension, comprises Dublin, Belfast, Cork, Queenstown, Galway, Drogheda, Dundalk, Athlone, Ballinasloe, Portadown, Maryborough, Thurles, Newry, Mullingar, Tipperary, and many other places, the Magnetic Company's lines in Ireland extending from north to south, and from east to west throughout the island", "Extension of the Magnetic Telegraph", *Dublin Evening Mail*, 13 January 1854.

[93] "The main line of the underground wires is laid under the mail road, passing from London through Birmingham to Manchester and Liverpool, and thence to all the principal towns of the Lancashire district. From Preston a line of six underground wires is carried through Lancaster, Carlisle, Dumfries, and Stranraer, to Portpatrick, at which point the wires are connected with the cable. From Carlisle a branch diverges to Edinburgh, Glasgow, Greenock, and other towns in the north of Scotland. The line from London through Lancashire to Scotland has been open for some weeks, and its efficiency and value are already well appreciated", Ibid.

Telegrams began to travel on the Irish circuit in early 1854, and the main Irish, Scottish and northern cities were interlinked with London, too.[94] Work went on in 1854 to install both underground and overhead links, improving the service between the cities already included in the network and extending it in Ireland and Scotland.[95]

By the beginning of 1855 the Magnetic stood out as the Electric's most frightening adversary. First of all, its corporate organization had been masterminded for a particular form of business; limited liability had allowed it to have a widespread shareholding and therefore a strong capitalization. To satisfy the local needs of the shareholders with activities spread nationwide, four boards

---

[94] "The underground trunk line of ten wires, from London to Manchester and Liverpool, was finished at the end of October, and has since continued in efficient operation. The underground line of six wires, from Liverpool to Carlisle, and thence to Portpatrick, was opened lately, and is working satisfactorily. The submarine cable of six wires, between Great Britain and Ireland was successfully laid from Portpatrick, in Wigtonshire, to Donaghadee, in May, and maintains the same perfect condition as I then reported to you. The underground line from Cork to Queenstown has been in operation since the commencement of the present year. The line of pole telegraph from Belfast to Dublin, with branches to Armagh, Kells, Newtownards, and Holywood, along the County Down, Ulster, Belfast Junction, and Dublin and Drogheda Railways, has been opened for commercial business in connexion with the cable since the beginning of January. The same system from London to Cork, with branches to Killarney and Carlow, on the Great Southern and Western and Killarney Junction Railways, was completed and opened for public use at the same time. ... Tenders have been accepted from responsible and experiences contractors for the erection of a line of telegraph upon the Waterford and Limerick Railway, and for six additional wires between Dublin and Belfast. Agreements of a satisfactory nature have also been entered into with the Londonderry and Enniskillen and Coleraine Railways for the supply of your telegraphs to those lines", Engineer's report in "Magnetic Telegraph Company", *Cork Examiner*, 6 March 1854.

[95] "Underground lines, consisting of six wires, have been laid between Dublin and Belfast and from Liverpool to Preston, and a line of four underground wires has been extended between Dumfries, Glasgow, and Greenock. These lengths constitute important links in the company's system, and by their completion direct communication has been established between Liverpool and Glasgow, Belfast or Dublin, in addition to the unbroken connection previously effected with the intermediate towns on these lines, and with London and Birmingham on the South. An overground line of four wires has been erected between Londonderry and Enniskillen and others are in course of formation that will connect Belfast with Ballymena, Coleraine, Londonderry, and Enniskillen. Dublin has been placed in communication with Limerick by means of the extension on the Waterford and Limerick Railway, and additional wires are being laid on the pole line south of Dublin to meet the increased traffic arising from the new line", "The Magnetic Telegraph Company", *Liverpool Mercury*, 16 February 1855.

of directors were created with seats in Liverpool, Manchester, London and Dublin. The executive committee, which controlled the overall network, met every day in Liverpool, where most of the shareholders were concentrated.[96] Secondly, the commercial and industrial nature of the shareholders had identified right from the beginning the potential target of users as businessmen. Unlike the Electric, whose exclusive concept of the service depended on the universalization of the postal service, the Magnetic repeatedly stated it was concentrating on the correspondence of the commercial and industrial classes, and thus avoided opening offices in unproductive areas.[97] Thirdly, when making its initial moves, the Magnetic had always shunned an aggressive approach with its rivals. For example, it had avoided locating offices in low-profile places where the Electric was already present,[98] and on many occasions, its managers declared they were against an unbridled competition, which they felt would harm all business.[99] And in fact the Magnetic had always preferred to secure the niches left vacant by the Electric, which guaranteed a relative monopoly. Then the Magnetic sought and partly managed to bypass the Electric in the exploitation of innovative technologies, which up to then had been the latter's reserve. Evidence of this is not only the introduction of the Henley telegraph and underground cables, but also the appointment of Charles Bright, who with his know-how on telegraph installations and his own packet of potential patents, functioned as a long-term investment.

---

[96] "As respected the charge to the directors, they should bear in mind there was a board in Liverpool, in London, in Dublin and in Manchester. The executive committee in Liverpool met every day", Ibid.

[97] "In the general extension of their lines, the directors had always been guided by the one principle, of only extending them to such places as they had every reason to believe would be remunerative, or to such places only as could give a guarantee that the receipts at least would be equal to the expenses", Ibid.

[98] A good example here is the case of Kendal. The Magnetic's line connecting London to Ireland passed through this little town, though the company avoided opening an office there because the Electric was already present in the railway station, "The Magnetic Telegraph Company", *Kendal Mercury*, 4 June 1853.

[99] "With respect to their rival companies, they were on terms of perfect amity; and they might depend upon it that so long, as the present directors managed their affairs, they would never be the first to enter into that ruinous system of competition which had been so very injurious to other public companies", "The Magnetic Telegraph Company", *Liverpool Mercury*, 16 February 1855.

3.4   The Submarine and the European: competition at sea

The origins of the third telegraph company to enter into competition with the Electric in the early 1850s go as far back as 1845. In the June of that year, Jacob Brett temporarily registered the General Oceanic Telegraphic Company, and a year later, together with his brother, John, officially registered the General Oceanic & Subterranean Electric Printing Telegraph Company, whose main aim was to establish a telegraphic communication from Great Britain and New Scotia in Canada. Two years later, John, who was not only the elder brother but also the more expert businessman, less ambitiously obtained a concession from the French Government to lay a telegraphic cable across the Channel. The agreement was renewed for ten years in 1849, but with the mandatory clause to establish communications before 1 September 1850. For this purpose the English Channel Submarine Telegraph Company was established, thanks to the personal subscriptions of John Brett, Charles Fox – later a shareholder of the Magnetic – Charlton Wollaston, engineer and ex-pupil of Brunel and Francis Edwards.

On 23 September 1850, after a month's work, the company managed to lay the cable, though it functioned correctly only for a few hours.[100] Some authors today think the worst and suspect that the insufficiently protected cable had been laid in haste in order to obtain the renewal of the concession.[101] The shareholders achieved their purpose, but at the same time the failure of the cable caused many possible investors to flee. The renewal of the agreement was due above all to the personal initiative of T.R. Crampton, an engineer and railway entrepreneur, who put down half of the capital needed for the laying of a second cable and convinced Lord

---

[100] Charles Bright, *Submarine Telegraphs. Their History, Construction, and Working*, London, Crosby Lockwood and Son, 1898, pp. 5–10.

[101] "The concession acquired by Jacob Brett in August 1849 required that electrical communication be made between England and France before September 1, 1850. To meet this deadline the Company laid a lightweight unarmoured, gutta-percha insulated single core cross channel cable on August 28, 1850. It was so light that it required lead weights to stop it floating. After a few messages the circuit failed, but the terms of the concession had been met and the Company gained time, and publicity, for the financing and manufacture of a much heavier, multi-core cable", Steven Roberts, *Distant Writing. A History of the Telegraph Companies in Britain between 1838 and 1868*, London, Steven Roberts, 2006–2012, p. 78.

de Mauley and James Carmichael to come in on the venture. At the beginning of 1852 these partners formalized their position with the establishment of a limited partnership, the Compagnie du Télégraphe Sous-marin, to safeguard the interests of the new partners and attract new capital. The company, then re-baptized the Submarine Telegraph Company with its official seat in Paris in order to maintain the French Government concession, set up an operative centre in London.[102]

Thanks to the capital and Crampton's technical supervision, the first really operative cable was laid across the Channel and opened a public correspondence service on 13 November. However, the Submarine only possessed the telegraph cable between Dover and Calais and had to use the South Eastern Railway Company to get its telegrams to London. In fact, following the establishment of the Electric, the South Eastern had requested that Cooke install the telegraph along its line in 1845 and had decided to run its own telegraph licence autonomously. Consequently, the South Eastern had adopted a very far-sighted policy from 1846 onwards and had defended its autonomy and exclusive rights over the London-Dover line.

With a possible success in laying the Channel cable in sight, in the spring of 1851 the Submarine's partners obtained parliamentary authorization to build telegraph landlines from Dover to London and other financial centres.[103] An autonomous line, possibly built along the same route as the South Eastern, would contain costs[104] and increase dividends. So with the positive results of the cable's first month's service, the Submarine partners advertised the sale of their shares, in view of a Royal Charter extending their powers and the shareholding of the European, a subsidiary of the Submarine. Lord De Mauley was chairman of both, while Arthur Anderson, founder and director of the Peninsular & Oriental

---

[102] Bright, *Submarine Telegraphs*, pp. 10–13.
[103] "Parliamentary Applications for the Ensuing Session", *London Daily News*, 2 January 1851; "Parliamentary Committees", *London Standard*, 3 June 1851.
[104] "The receipts of the last month yielded a return of nearly 8 per cent, on the total of capital and expenses on the part completed between Dover and Calais; the amount paid the South Eastern Railway Company from Dover to London being 16 per cent on a capital sufficient to supply four times the amount of accommodation from Dover to the Offices at Cornhill", "Instantaneous communication between Liverpool and the Continent of Europe by the European and Submarine Telegraph Company", *Liverpool Mercury*, 7 May 1852.

Company, John Brett, James Carmichael, W.J. Chaplin, chairman of the London & South Western Railway Company, John Masterman, Samuel Laing and Ernest Bunsen were all directors of both companies.[105] Two years later the two latter partners bought a substantial block of shares in the British.

When faced with the impossibility of placing another line along the South Eastern railroad,[106] the European began work on installing an underground cable along the old carriage road.[107] Lagged in gutta-percha, it contained six copper wires and was inserted in a trough, then placed in a trench dug along the road.[108] The construction technique was similar but less refined than Charles Bright's used for the British and the Magnetic. Thanks to the surprisingly rapid installation, the Dover-London line was rapidly completed and officially inaugurated on 1 November.[109] Within a few weeks the Submarine/European was able to start off a direct

---

[105] "The Submarine Telegraph Company (Advertisement)" and "The European (and American) Electric Telegraph Company", *Morning Post*, 1 May 1852.

[106] "The European Telegraph Company and the South-Eastern Railway not having been able to come to terms for erecting additional wires from London to Dover, the Telegraph Company have resolved to effect a communication along the old high-road to Dover. The wires, covered with gutta percha, are laid in a trough, which is buried in a trench; the work is completed from London to Chatman, and a mile and a half of additional wires are laid down each day", *Devizes and Wiltshire Gazette*, 23 September 1852.

[107] "In a few weeks will be completed a second line of electric communications, in connection with the continental telegraph, between Dover and the metropolis. It has been promoted by the European Telegraph Company, and one of its peculiar novelties is that it is being laid down along the old coach road through Deptford, Greenwich, Shooters-hill, Dartford, Gravesend, Strood, Rochester, Chatman, Sittingbourne, Faversham, Canterbury, etc., to Dover", "Subterranean line of Electric Telegraph between London and Dover", *Reynolds's Newspaper*, 19 September 1852.

[108] "The copper wires, six in number, are encased in gutta percha; and being deposited in a kind of trough, constructed of kyanised timber, it is laid in a trench dug in the road, some foot and a half from the surface. In order that there should not be the possibility of the wires failing, test boxes, by which the wires are proved, are erected every mile. The works are proceeding with the utmost expedition. A mile and a half is completed every day", "Laying down of a subterranean line of electric telegraph between London and Dover in connection with the submarine cable", *Leeds Mercury*, 25 September 1852.

[109] "Today this telegraph, which has been laid along the old coach road to crosses the Channel to Calais, is to be thrown open for public use, and will give a second line of communication with the metropolis and the Continental telegraph. It is the property of the European Telegraph Company. It is quite independent of the South Eastern Railway Company and will be shorter and cheaper", "Opening of the Subterranean telegraph to Dover along the old Coach Road", *London Daily News*, 1 November 1852.

telegraph service between London and Paris, with a soaring increase in revenues.[110]

With the enormously successful activation of its first line, the European started building links from the capital to Liverpool, Manchester and Birmingham in 1853,[111] in open competition with the Magnetic, which had only just begun its service. The European had, however, a supreme advantage over all the others in that it communicated directly with the continent, which also led it to further extend its network within London and link up with the cities in the North.[112] At the beginning of 1854, the European appeared to be in excellent health, as witnessed by the positive data unveiled at the shareholders' meeting and above all by the dividends paid out.[113]

----

[110] "Towards the close of last week the new line of telegraphic wires lately being laid down in troughs over the old London road to Dover, and so connecting the submarine telegraph between Dover and Calais, which again is directly attached to the line of telegraph between Calais and Paris, with the batteries of the Submarine Telegraph Company, whose offices are in Cornhill, was completed, and on Monday, for the first time, messages were exchanged direct between London and Paris, the messages being transmitted from capital to capital instantaneously, and as instantaneously being replied to. The directors of the English public company to whom Europe is indebted for the consummation of the submarine system, seem to have taken a proper view of the importance of the work they had finished, in calling together at their offices in the City the large and eminent party of gentlemen – commercial magnates, railways notabilities, and distinguished foreigners, proximately and indirectly representing other nations – met on the occasion to 'assist' in and celebrate the great fact", "Direct telegraphic communication between London and Paris", *Kentish Gazette*, 9 November 1852.

[111] "The direct electric wires, to connect these important cities with Continental Europe, *via* France and Belgium, will commence to be laid down to day (Feb.1) by the Submarine and European Telegraph Company", "Liverpool, Manchester, Birmingham, and the Continent", *Morning Post*, 1 February 1853.

[112] "Yesterday the Submarine and European Telegraph Company commenced laying down their wires in Pall-Mall, St. James's-street, and Charing-cross, in order to leas branches into the principal club-houses at the West-end, the government offices, Admiralty, Houses of Parliament, and Buckingham Palace, so that instant and direct communication may be made, without dispatching messengers to the central office, with all parts of the European Continent reached by electric telegraph", "Extension of the Submarine Telegraph", *London Standard*, 7 April 1853.

[113] "The number of messages transmitted during the month of January, 1852, after the opening of the French line, was 1,068; in January 1853, 2,018; and in January, 1854, the number had increased to 3,120. This increase has been maintained in spite of all commercial affairs from the unsettled aspect of the political horizon. The present position of the company's finances enables the directors to recommend a dividend at the rate of 8 per cent per annum for the last six months, on the capital of 75,000 l", "The Submarine and European Telegraph Company", *Dublin Evening Mail*, 10 February 1854.

Nevertheless, in the August of the same year the British announced its coming takeover of the European.[114] It was buying all the installations, offices and staff of the European, offering shareholders £100,000 in shares and £30,000 in cash. Moreover, the 11 directors of the European were automatically brought onto the board of the British.[115] The operation was fully understandable from the point of view of the British, which in one stroke acquired its much-desired link with London, a complete network already operating in the North and the exclusive right to send telegrams via the Dover-Calais cable.[116] More uncertain are the motivations of the European. Most probably, its shareholders – first movers in the field of submarine telegraphs – had no desire to be caught up directly in the difficult land competition with the Electric. Then it obtained a substantial economic compensation and the privilege of a position in the governing body of the British. Last but not least is the detail that in October 1854, a significant number of British shares passed to a consortium of three shareholders, directors of both the Submarine and European.[117] Though sources up to now provide no backing, it could be legitimately thought that the proprietors of the

---

[114] "The British Telegraph Company has completed an arrangement with the European Telegraph Company, by which the two undertakings become one, under the title of the former, by the purchase of the property of every description belonging to the latter company. The purchase has been effected by the payment of L100,000 in the shares of the British Telegraph Company, and 30,000 L. in cash. By this arrangement a more complete system of communication is established", *Bradford Observer*, 10 August 1854.

[115] "It has been agreed that eleven directors of the European Company shall be elected directors of this company; and, in pursuance of that arrangement, the board have to recommend to the shareholders at this meeting the appointment of the following gentlemen as an addition to the present board, viz: – Lord de Mauley, Samuel Laing, Esq., M.P.; W. J. Chaplin, Esq., M. P.; the Hon. F. W. Cadogan; Sir James Carmichael, Bart, Ernest Bunsen, Esq.; Rear-Admiral Sir Richard O'Conner; Robert Gill, Esq.; and George Edward Dering, Esq", "British Telegraph Company", *London Daily News*, 4 September 1854.

[116] "The attention of the directors was next anxiously turned to the completion of the Communications to London, in the speediest and most effectual manner. At this juncture the European and American Electric Printing Telegraph Company had just completed the laying down of their wires from Dover to London, and thence to Liverpool and Manchester and a correspondence having been opened with that company, they were found willing to unite their lines with those of the British Company, thus in connexion with their own northern lines, supplying to the latter, without the delay of construction, the communication from Liverpool, Manchester, Glasgow, Hull, Newcastle, and intermediate stations to London, and at the same time placing those districts in direct communication with the continent of Europe", Ibid.

[117] BT, *British and Irish Telegraph Company*, TGL 1/1/1, *Shareholder List*, 1854.

Submarine preferred to relinquish the direct management of the land telegraph to a company reputed competitive in this specific market, with the dual advantage of a limited responsibility in the venture and the immediate acquisition of a potentially enormous catchment area for telegrams to/from Europe.

3.5 The first mover supreme

During the early 1850s, i.e., as long as Ricardo managed to keep the control of the company in his hands, the Electric's strategy was to consolidate its dominant position in the market. This can be seen in the constant expansion of its correspondence activity, its way of developing its network and its desire to respond blow for blow to the technological and commercial offensives of its rivals.

Under Ricardo's leadership, telegraphic correspondence and management of intelligence had become the Electric's core activities, inevitably imposing a long-term infrastructural investment. The same policy was followed when rivals came into the market, as is witnessed by the unrelenting increase in mileage of lines constructed, telegrams transmitted and takings.

The data available not only witnesses the constant quantitative growth of the network but also the impressive acceleration of the correspondence. In fact, while the lines were tripled and the miles of wire quadrupled, the number of messages increased more than sevenfold. That meant that once the first block of heavy investments had been effected between 1848 and 1851, the correspondence activity increased more than proportionally with the increase in mileage of lines. According to Ricardo, a telegraph wire could become fully productive only some years after its installation, when new users had been won over and become familiar with the service. From this point of view, a constantly expanding network would allow the Electric to be always one step ahead of its rivals.[118]

---

[118] "I wish you to understand clearly, that no Telegraph established in a town at once gets into successful operation. It takes some time, some considerable period in many towns, though it is as it were the addition of a new sense to the public, to give them facilities of instantaneous communication with other places, it is a long time before they make their commercial arrangements and so combine their transactions as to make the Telegraph available for the purposes of business", BT, *Electric Telegraph Company*, TGA 2/1, *Speech of J.L. Ricardo, Esq., M.P., Chairman at the General Meeting, held January 29, 1855.*

Table 3.1   Rate of development of the Electric's network

	Lines (miles)	Wires (miles)	Messages	Receipts (£)
June 1850	1,684	6,730	29,245	20,436
December 1850	1,786	7,200	37,389	23,087
June 1851	1,965	7,900	47,259	25,529
December 1851	2,122	10,650	53,957	24,336
June 1852	2,502	12,500	87,150	27,437
December 1852	3,709	19,560	127,987	40,087
June 1853	4,008	20,800	138,060	47,265
December 1853	4,409	24,340	212,440	56,919

Source: BT, *Electric Telegraph Company*, TGA 2/1, *International Telegraph Company* (*incorporation*).

At the beginning of 1855, the Electric could therefore count on almost 5,000 miles of lines against the 2,000 of the Magnetic, the rival with the most extended network.[119] Unfortunately, the overall length of the lines and wires is only an indicator of the potentiality of a communication infrastructure and does not allow us to appreciate the grade of development of the traffic. To measure the latter, we need to know the overall number of telegrams sent, data furnished only by the Electric and only for a few specific years.[120]

As first mover the Electric had originally planned the development of its network over the whole territory, connecting all the main commercial centres of the British Isles. Given the presence of the Penny Post, its original concept was of an exclusive service but with a national coverage.[121] The arrival of competitors who exploited

---

[119] "The total mileage of the Company [Magnetic] is therefore a little above 2,000 miles, and the length of wire about 13,000", Charles Bright, answer n.1, Appendix IV, Edward and Charles Bright, *The Life Story*, pp. 424–425.

[120] The distance between the Electric and the Magnetic would appear even greater if we had data on their telegram traffic. But we only possess financial data, which tells little of traffic figures for two reasons. There can be takings coming from other sources, and the companies used different tariff policies, so that similar sums recorded can represent extremely different numbers of telegrams.

[121] "From the time that the Directors, in accordance with the policy announced at a former meeting, aimed at the establishment of a great national system of Telegraphic Communications, rather than the large manufacturing profit, which the exclusive nature of their Patents might have secured to them, their attention has been anxiously turned to the development and extension of their Lines, and they have thought themselves justified in sparing no expense to render the accommodation comprehensive, certain, and expeditious. ... but your directors will not rest satisfied till all parts of the

niches of the market it had left vacant emphasized its national image, which influenced possible users. This can be seen in the fact that despite the press's strong opposition, the Electric's mostly traditional clients rarely turned to other companies.[122] In the same way, the cities wanting to open a telegraph office often went on using the Electric precisely because it possessed a nationwide network.[123] And for its part, Electric opened offices wherever there were lobbies ready to finance the venture and buy the necessary shares.[124]

Nevertheless, it was in the technological field that the Electric carried out its strongest counteroffensive. From 1854 onwards, an ever increasing interest in improving line insulation began to emerge from the periodical meetings held by the Electric's superintendents. Their in-depth investigations revealed that the sulphur used to protect the iron components from the weather caused the frequent

---

Kingdom shall be provided with Telegraphic accommodation, and until they have so combined and economised the system as to bring it into the most extended use for which it is naturally adapted. The Company are prepared to establish a Station at any point traversed by their wires, upon their working expenses as such Station being guaranteed to them", BT, *Electric Telegraph Company*, TGA 2/1, *Directors' Report June 1851.*

[122] "Whatever customers have tried the 'Magnetic Telegraph' there have not been any instances when those who have been experienced the manner in which the business has been conducted have discontinued the use of this Company's Telegraphs from a preference for the old network", BT, *English and Irish Magnetic Telegraph Company*, TGN 1/1/1, *Memoranda relating to General Meeting*, 4 January 1853.

[123] "The Telegraph Monopoly", *Newcastle Journal*, 2 July 1853.

[124] "For instance, a place like – say King's Lynn – comes to us and says, 'You have got your Telegraph within fourteen miles of us: will you extend your wires to our town?' They show you that they have a traffic which will pay a larger profit than the average profit you are obtaining from the whole of your undertaking; and they say, 'We shall be ready to give you all the assistance and all the facilities we can if you will bring your Telegraph into our town.' Now, we have only one way of meeting such applications; and I want to call your particular attention to this subject, in order that you may understand a letter which the Directors will think it necessary to send to the Proprietors. It being clearly proved that an extension such as this will be a profitable piece of Line, that it will be a great public convenience, and it being very difficult for a Company, which is almost a national institution, to refuse to give all the facilities and accommodation in its power to the public, we tell them this, – 'If you will take shares at par equal to the amount which will be necessary for its construction, we will put up the Line for you; but we cannot raise capital, because our shares are at par, and we do not intend to borrow it by debentures or otherwise; but if you take our shares at par, we shall then be willing to employ the capital so raised in making your Line'", BT, *Electric Telegraph Company*, TGA 2/1, *Speech of J.L. Ricardo, Esq., M.P., Chairman at the General Meeting, held January 2, 1855*, p. 4.

cracking of glass insulators along the lines.[125] A long debate therefore followed to assess the possibility of introducing new style
insulators made of different materials.[126] Some months later, the
discussion was over replacing glass with earthenware.[127] Then the
superintendents examined the possibility of improving the quality
of the lines by injecting tar and other chemical substances into
the wooden poles to increase their weatherability.[128] In the same

---

[125] "Unanimously recommended that Sulphur be abandoned", BT, *Electric Telegraph Company*, TGA 1/7/1, *Quarterly meetings of Superintendents Reports*, 3 June 1854, p. 22.

[126] "Mr. Ansell reported upon the result of the experiment upon one of the Admiralty wires between London and Bristol. No benefit seems to result from the adoption of this plan, and it was recommended that these experiments be abandoned. A Glass insu-lator schemed by Mr. Britton, was exhibited, but it was considered that it involved no improvement in principle or design, superior to our present insulator, and therefore, its adoption was not recommended. Another insulator (an adaptation of Walker's cones) was also exhibited, but for the above, and various other reasons, it was considered objectionable. This insulator was a production of Mr. Culley's", BT, *Electric Telegraph Company*, TGA 1/7/1, *Quarterly meetings of Superintendents Reports*, 4 October 1854, pp. 14–16. "The question of the best form of insulator for lines exposed to salt spray was discussed. Mr. Walker's insulators were considered applicable in such cases. A disk or wheel made entirely of glass or earthenware, attached vertically by its circumference to the arm, with wire passing through the centre, was suggested by Mr. Walker. Mr. Culley also suggested a disk, but placed horizontally. The first was considered the most practicable and best, because it allowed both sides to be washed by the rain", BT, *Electric Telegraph Company*, TGA 1/7/1, *Quarterly meetings of Superintendents Reports*, 5 December 1854, pp. 10–11.

[127] "The question of the best form of insulator for lines exposed to salt spray was discussed. Mr. Walker's insulators were considered applicable in such cases. A disk or wheel, made entirely of glass or earthenware, attached vertically by its circumference to the arm, with wire passing through the centre, was suggested by Mr. Walker. Mr. Culley also suggested a disk, but placed horizontally. The first was considered the most practicable and best, because it allowed both sides to be washed by the rain....A glass insulator, having the bolt passing entirely through, with the head countersunk in the insulator, and covered with a plug, cemented in, was exhibited. This form was univer-sally approved, and ordered by Mr. Clark to be practically manufactured. The advis-ability of abandoning GLASS for EARTHENWARE was discussed. It was considered that glass was undoubtedly superior to earthenware as an insulator, but great inconvenience had been experienced by its introduction, from the numerous changes that had been made. It was also thought that glass had hardly had fair test at present", BT, *Electric Telegraph Company*, TGA 1/7/1, *Quarterly meetings of Superintendents Reports*, 5 December 1854, pp. 10–11. "A new earthenware insulator, shaped like the present glass, to supersede the old earthenware and zinc, was submitted and much approved of", BT, *Electric Telegraph Company*, TGA 1/7/1, *Quarterly meetings of Superintendents Reports*, 6 April 1855, p. 12.

[128] "He [Mr. Edwards] also suggested the advantage of tarring the poles altogether in place of painting them, as being both better and cheaper; but it was considered that the unsightliness of this plan, and the consequent objection which Railway Companies would have to its use rendered its adoption objectionable. ... He also enquired whether

meetings they constantly discussed introducing innovations which would reduce the cost of the batteries.[129]

Having been pre-empted by the Submarine in laying the first Channel cable, in 1852 the Electric obtained from the Dutch Government authorization to put down and run a submarine cable between Great Britain and Holland.[130] Permission was granted for placing the cable but not for direct management. The Electric reacted by creating the following year a subsidiary company named the International Telegraph Company,[131] which thanks to a Royal

---

tarring the pole from the top of the pole to the bottom arm would not be beneficial. This question was discussed and it was recommended that if Mr. Edwards could obtain leave from Mr. Seymour Clark, the experiment should be tried between London and Peterborough, and that the result of this experiment, and of that of tarring the pole caps, should be read at a future meeting", BT, *Electric Telegraph Company*, TGA 1/7/1, *Quarterly meetings of Superintendents Reports*, 4 October 1854, pp. 22–23; "Mr. Clark brought to the notice of the meeting a system of preserving wood, the invention of a Frenchman, which had been largely introduced in France", BT, *Electric Telegraph Company*, TGA 1/7/1, *Quarterly meetings of Superintendents Reports*, 7 July 1855, p. 9.

[129] "The system of working several circuits from one battery was brought forward and fully discussed. Mr. Edwards reported very favourably upon the results of their adoption at Peterborough. Mr. Culley had also tried the plan with much success at Manchester and Stoke, with resistance coils. The plan adopted at Strand, and that recommended for general adoption, was that of having divided batteries, viz., each half being placed to earth. The attention of the Superintendents was particularly called to this system, which tended so greatly to economise our battery department. ... Mr. Ansell reported upon some experiments he had made upon Callan's Single Fluid Battery. He had found it very effective. He used cast iron and zinc, with sand moistened with a solution of one part sulphuric acid, and three parts common salt. He had found the power very constant, and more intense than our common sand battery. He had one which had been in use for a month, and which had only refreshed twice. The chief advantage in this battery was, that it would enable us to use our guttapercha troughs", BT, *Electric Telegraph Company*, TGA 1/7/1, *Quarterly meetings of Superintendents Reports*, 6 April 1855, pp. 10–11.

[130] BT, *International Telegraph* Company, TGC 1/1, *Proposals for a submarine cable between the UK and the Netherlands.*

[131] "I want now to say a Word about that International Company. I believe it will be in the recollection of all those Proprietors who held Shares at that time, that in the year 1852 we obtained the concession from the Dutch Government, and that this being secured, a vote was passed at a General Meeting of our Proprietors that we should advance money to the International Company, in order to take advantage of the concession, by which it was stipulated that the International Line to the Hague should be completed within a certain time. At that period we should willingly have disbursed the sum necessary for the laying down of the cable from our funds, making the work simply an extension of our own system. We had the means and the power by our Act of Parliament of doing this; but the reason why we refrained from taking this course was, that the Dutch Government objected to the Line in Holland being made by the English Electric Telegraph Company; and, in order to get over that difficulty, we made

Charter enjoyed the privilege of limited liability.[132] Its capital was composed of a third of the shareholders of the Electric, who had put down a sum equivalent to the costs of the cable, a third of the shares sold on the London market, and a last third sold in Holland.[133] The International's board actually corresponded to the Electric's, while the management was entrusted Douglas Pitt Gamble as secretary and Edwin Clark as chief engineer, two of the Electric's men.[134] Between May and September 1853,[135] the International laid and made operational three different submarine cables, each containing a single conductor. Needless to say, once the International's cables reached Dover, they were connected up to the Electric's land network.[136] Back in November 1852, the Electric had acquired the landing rights of the Irish Submarine Telegraph Company, which had failed to lay a cable between Ireland and Scotland.[137] Some years later, in September 1854, the International exploited the acquired rights and placed a second cable between the two islands. Finally, in 1855 a parliamentary authorization allowed the International to be absorbed by the Electric.[138] In this way, the Electric had managed in little more than three years to activate international connections with Europe as well as Ireland, and close the small technological gap created by its rivals who had laid the first British cables.

---

a fresh Company, though we had the same Directors and the same staff: in fact, we had almost the same elements to go upon, and the same capital, I may say, yet we ostensibly made a fresh Company; and while we received authority from the Dutch Government, on the one hand, we obtained, on the other, a charter from the Board of Trade here to enable us to form the Company in this country", BT, *Electric Telegraph Company*, TGA 2/1, *Speech of J.L. Ricardo, Esq., M.P., Chairman at the General Meeting, held January 29, 1855*, p. 3.

[132] BT, *International Telegraph* Company, TGC 1/2, *Royal charter to form a company for laying submarine cable to Holland*.

[133] BT, *Electric Telegraph Company*, TGA 2/1, *International Telegraph Company* (proposal of incorporation).

[134] Edwin Clark was at the time chief engineer, having taken over from Hatcher in 1850. His brother, Latimer, was assistant engineer and would take his place as chief engineer of the Electric in 1854.

[135] "Electric Telegraph from England to Holland", *Morning Chronicle*, 6 June 1853.

[136] BT, *International Telegraph Company*, TGC 1/4, *Board Meeting Books*.

[137] "An amalgamation between the Electric Telegraph Company and the Irish Submarine Telegraph Company, recently incorporated by Royal Charter, is being carried out for effecting this object", "Submarine telegraph between Holyhead and Dublin", *Hampshire Advertiser*, 25 September 1852.

[138] *An Act to consolidate the Capital Stock of the Electric Telegraph Company and of the International Telegraph Company, and to grant further Powers to the Electric Telegraph Company* [2 July 1850], 18–19 Victoria, c. cxxiii.

Between 1853 and 1855 the Electric managed to remodel its ownership to make itself more competitive with the Magnetic, which,
thanks to a broad-based shareholding and limited liability, appeared
readier to face the new technological challenges of the market. In
1853 the Electric was authorized to increase its capital,[139] and in the
following year limited liability for its shareholders.[140] The effect on
the sale of shares was immediate, and by the end of 1855 the Electric's
dividends were distributed among 400 shareholders.[141]

Paradoxically, the strategy of defending the Electric's monopoly
conducted by Ricardo led progressively to the entry of new partners,
who were more in favour of immediate returns than future investment, so determining the gradual change in the company policy,
which was confirmed by Ricardo stepping down from the chairmanship and the board.

3.6   The consequences of the constitutive choices

The choices made before 1850 had forged the core business of telegraphs, throwing up three main categories of users: the railways, the
press and businessmen, categories which determined in a marked
way the dynamics of rivalry in the early 1850s.

First of all, the railway companies generated two opposing forces:
on one side the Electric, defending its position of dominance, and on
the other new rivals and their supporters. The compact defence the
railway companies gave the Electric during the British's attempted
to reach London and the ensuing parliamentary battle was due to
the fact that the railways were both users and shareholders of the
Electric. The contribution of the railways was also essential for
the development of the Magnetic, the British and the Submarine/
European. Both the British and the Magnetic built their telegraph
lines where the Electric was absent, and all three rivals of the Electric
were run by top men closely linked to the railway world and had
many railwaymen among their shareholders.

---

[139] *An Act for granting further Powers to "The Electric Telegraph Company" and to enable
such Company to make Arrangements for the working of Telegraphs adjoining their Works* [15
August 1853], 16–17 Victoria, c. cciii.
[140] *An Act for limiting the Liability of the Shareholders in the Electric Telegraph Company, and
for granting additional Powers to such Company* [31 July 1854], 17–18 Victoria, c. cciii.
[141] BT, *Electric Telegraph Company*, TGA 1/6, *Shareholders Books*.

The press, however, encouraged a strong anti-monopolistic current favouring the British, but also offered considerable help to the Magnetic during its period of greater development between 1853 and 1855. The press was always compact in its attacks on the Electric and its policy of news distribution, with a tendency that grew stronger as time passed and reached its highest point during the process of nationalization between 1865 and 1869. In their turn the businessmen, too, very interested in the rapid flow of news, upheld the Electric's main rivals, both of which based their capital structure on a diffused shareholding representing businessmen in commercial and industrial cities, particularly Liverpool, Manchester, Birmingham and Glasgow.

Government and Parliament apparently remained on the margins of the dynamics of the telegraph market in the 1850s. But a series of provisions preventing overhead lines being built within towns and cities gave an even greater thrust to research on covering the cables with gutta-percha. It was no coincidence that many of the engineers who installed underground lines for the Magnetic, British and European later became leading figures in the great submarine telegraph ventures.

4
The Duopoly

The telegraph market underwent a radical change during the late 1850s, when it passed from a state of quasi-free competition with warring companies to a duopoly held in place by more or less official cartel agreements. The turning point was the year 1855, when events both exogenous and endogenous forced the managerial class to take radical decisions in sharp contrast with their usual strategies. Thus they favoured the rise of new long-term characteristics in order to favour small shareholders after high dividends. Three events in particular stand out: the reinforcement of a new majority on the Electric's board; the merger between the Magnetic and the British leading to the birth of an alternative group; the duopoly created between the Electric/British & Irish.

Although the consolidation of the duopoly was financially rewarding for the shareholders, it generated a series of negative externalities which led to customer disaffection. And that was the fertile breeding ground for the growth of a hard core of criticism, mostly from the press and businessmen, calling for radical changes in the sector.

4.1 The turning point

In 1855 four closely linked events took place which were destined to make a drastic change in the balance of power in the sector: a price agreement between the three main companies; changes at the top of the Electric; the technical crisis affecting underground lines; preliminary negotiations for the construction of the first trans-atlantic cable.

Taking up the spirit of competition aroused in the second half of 1854, the Submarine, whose majority partners also held a substantial packet in the British, started a price war with their competitors and thereby stirred up much feeling in the public. Tariffs were slashed by 50%, while the Electric and the Magnetic reluctantly followed suit in an attempt not to lose ground.[1] While the events of the Crimean War naturally sent up the number on dispatches on state affairs, there was a sharp drop in the core activity of commercial and bourse telegrams, slowed down like all economic activity by the war.[2] The lack of reaction in users to such reductions in tariffs and the trifling increase in returns led the companies to make an agreement in which a higher common price was fixed. The news of this cartel was first given by the press with a certain disappointment,[3] then described in the Directors' Report of the British and also mentioned in the

---

[1] "There was another circumstance which had tended considerably to diminish the receipts, and that was the reduction of the rate of charges to about 50 per cent...This heavy reduction was first made by the Submarine Company, when they extended their wires here. The Electric Telegraph Company followed, and this company endeavoured to battle through, but was at length obliged to yield", from the Directors' Report, "The Magnetic Telegraph Company", *Liverpool Mercury*, 16 February 1855.

[2] Ricardo explains it very clearly during his speech to the shareholders' meeting of January 1855: "It is all very well to make arrangements for long distances of Telegraph, to bring news from the seat of war; but what we look to is, the steady receipts from Commercial Towns. We look to our Commercial Business. We do not want the fitful and temporary excitement arising from the war. That is not the business which is profitable to us. What is profitable to us is steady Commercial Business; and if I were to pick from the Map of Europe four places to join in direct Telegraph Communication with each other for their own benefit, and for the advantage of this Company, it would be the four great Commercial Cities of Hamburg, Amsterdam, London, and Liverpool.... I have also to say, that we have had this Half-year several disadvantages to contend with. In the first place, we have had that unfortunate epidemic which raged during the summer months of the years, and which, as I think you must be all aware, put stop for a certain time to almost all business transactions. But we have had also, and we have it now more particularly, to contend with, that which I think we must look upon as materially influencing every class of business and all sorts of commercial transactions in this country. I mean the War in which we are at present engaged", BT, *Electric Telegraph Company*, TGA 2/1, *Speech of J.L. Ricardo, Esq., M.P., Chairman at the General Meeting, held 29 January 1855*.

[3] "We understand that the various Electric Telegraph Companies raised their fares yesterday. The announcement of the fact in the Stock Exchange, where business is at present the exception instead of the rule, was received with decided marks of disapprobation", *Westmorland Gazette*, 18 August 1855.

shareholders' meeting, held in September 1855.[4] Unfortunately, no official company document speaks clearly of such an agreement, so that the reasons behind it remain unknown.[5]

From what happened before and afterwards, some reasonable assumptions can be made. Barton, for example, the only scholar to indepth the subject, holds that it was the upsizing of the rival Magnetic and British which brought the Electric to give its consensus.[6] He also holds that the agreement followed on from the change of power on the board of the Electric.[7]

The absence of the board's minutes make it difficult to establish if the cartel agreement was the result of the changed majority or if it determined Ricardo's leaving the scene as a result of his loss of command. What is clear is that at the turn of 1855 there was a troubled transition in the leadership of the Electric which would have a

---

[4] "The directors, in common with those of the Electric and Magnetic companies, have agreed upon a reconstruction of prices, on a principle which they consider to be at once fair towards the public, and likely to prove justly remunerative to their shareholders. The new tariff began to take effect on the 1st August, and up to this time the directors have every reason to feel generally satisfied with the change", "British Telegraph Company", *London Daily News*, 4 September 1855.

[5] The BT archives conserve all the extant documents of the landline telegraph companies. While the Magnetic and British documentation is fragmentary and scarce, the Electric offers complete series from its beginnings to nationalization, with many kinds of sources. However neither the minutes for the board or shareholders' meetings for 1855 are available. The only primary source which mentions in passing a possible agreement is an entry in the Chairman's Book, which records all in correspondence to the Chairman, then Ricardo. "Magnetic Tel. Co. – Mr. Dunlop's remarks relative to coming to some arrangement with this Co., to avoid competition", BT, *Electric Telegraph Company*, TGA 1/1 *Chairman's Books*, n. 276, 24 April 1855.

[6] "But in July 1855 the English & Irish Magnetic Telegraph Company had Brown to the point where it was launching its national service and becoming significant. ... Should ETC expand its geographical footprint and increase its service offering, or cut its prices? For the previous few years it had done both including continuously sharply cutting prices, although not that for intelligence. Following the Submarine Telegraph Co. successfully crossing the Channel, it too also laid submarine cables to Holland. These actions cost substantial new capital, much of which was raised from new shareholders in Manchester. The price cuts led to substantial volume increases although initially not enough to prevent a fall in profits. ... In addition to the establishment of the intelligence cartel the telegraph companies organized a price cartel, starting on 1st August 1855, amounting to a 15% price increase", Roger Neil Barton, "New Media. The birth of telegraphic news in Britain 1847–68", *Media History*, 16/4, 2010, pp. 379–406.

[7] "A possible interpretation of this change of policy is that the composition of the boardroom had significantly changed leading to Lewis Ricardo losing control of it. Quite a few new ETC directors were appointed during the mid-1850s which changed the balance of views towards the economy", Ibid., p. 391.

long-term influence on its business strategies. From June to December 1855, the board increased considerably in size, passing from 10 to 16 members, mainly representatives of the provincial shareholders' committees – three from Manchester (Robert Chadwick, Edward Riley Langworthy and Joseph Whitworth), one from Liverpool (Brown), and one from Edinburgh (Lord Berriedale, also known as the Earl of Caithness).[8]

Splitting up capital to such a level had favoured the birth of syndicates of businessmen working in the same city and sharing the same desire for fat earnings over a short period. Unlike the historical shareholders of the Electric, such as Ricardo, Cooke, Bidder and Brassey (who had kept on the board his two trusted partners, Morton and Till),[9] the newcomers wanted to avoid reinvesting returns over the long period and preferred them to be distributed as dividends. In this way the first mover would be disadvantaged over the long period, though a price agreement would guarantee high returns in the immediate future. Barton claims that a helping hand in this direction came with the entry of William Henry Smith Junior,[10] the proprietor of a news agency of the same name. Though he only possessed a few shares, Smith was on the board by 1853. He was a stalwart conservative, an opponent of free trade, and on more than one occasion had operated an absolute monopoly in his field[11] and as such became Ricardo's strongest political adversary on the board. Seeing the defeat of his policies, Ricardo resigned during the shareholders' meeting of January 1856 and passed the chairmanship over to Wylde, his vice-chairman.[12] The choice was obviously a safe one, given Wylde's establishment contacts and closeness to the centres of powers.

---

[8] BT, *Electric Telegraph Company*, TGA 2/1, *Balance Sheet June 1855* and *Balance Sheet December 1855*.

[9] BT, *Electric Telegraph Company*, TGA 1/6, *Shareholders' books*.

[10] Barton, *New Media*, p. 391.

[11] Herbert Maxwell, *Life of the Right Honourable William Henry Smith M.P.*, London, William Blackwood and Sons, 1894; Charles Wilson, *First with the News: The history of W.H. Smith, 1792–1972*, London, Jonathan Cape, 1985.

[12] "Major-General Wylde, C.B., in the chair. ... The Chairman laid upon the table the balance-sheet and statement of receipts and expenditure of the company for the half year ending 31st December. He regretted to say that occupied the chair that day, in consequence of the resignation of their late chairman, Mr. Ricardo. ... Mr. Ricardo, after expressing his acknowledgments, said that whatever labours he had gone through were amply recompensed by so gratifying a compliment. That those labours had been great he must admit, and at one time so extreme were the difficulties that he almost

Barton's reading of events does not, however, take into consideration another possibility, that the other companies, too, particularly
the Magnetic, were also in hot water and that the cartel agreement
was a perfect compromise for all involved.

In March 1855 a first alarm bell was rung in a letter to the editor
published in the *Cork Examiner*. The Magnetic's superintendent
for Ireland gave notice that the delays and hardships the users
were experiencing originated in a not identified breakdown of the
lines for which a solution had not yet been found.[13] He was probably referring to the gutta-percha, which had given way along all
the Magnetic's underground lines in the space of a few months. As
Edward Bright later reported, the gutta-percha inserted in the metal/
wooden tubes had dried out damaging the cable lagging and its
capacity for insulation. In 1856 Charles Bright, the main believer
in the Magnetic's underground cable network, attempted to retrieve
the insulating capacity of the gutta-percha with a machine injecting
liquid under the lagging. The experiment did not work and Bright
was not able to bring the lines back into service. Foreseeing the
collapse of the whole network, the Magnetic's management started
a series of negotiations at the end of 1856 with rival companies
and decided in the end to merge with the British.[14] The collapse

---

despaired of success on bringing the telegraph to its present proud position, indeed at
one period there were but ten shareholders left, and they had bought up all the shares
that the holders were endeavouring to get rid of. They were then losing £ 300 per week.
In seeking rest from the labours of his Chairmanship, he was gratified to think that he
left the company in the best possible position. Their prosperity would go on increasing,
because the telegraph was only in its infancy", extract from the minutes of the *Electric*
shareholders' meeting in "The Electric and International Telegraph Company", *Morning
Chronicle*, 31 January 1856.

[13] "There is no doubt that the information has been irregular, and much delayed. This
arises from the frequent stoppages produced by the imperfect insulation of the lines.
Extensive alterations and improvements are being made, in fact the entire renewal of
the insulation of four old, and the addition of two new wires between Dublin, Cork,
Waterford and Limerick. Under the most favourable circumstances this is a work of
time, and of such a nature that none but those who have studied the subject can be
aware of the difficulties attending the execution. ... During the past frost it has been
almost impossible to make any progress. The weather is now improving, and every exertion will be made to recover the lost time, and I hope by the end of this month a line
will be completed that, as far as the science and money can do so, will secure a certain
and sufficient means for every purpose of private or public telegraphic communication", "To the editor of the Cork Examiner", *Cork Examiner*, 2 March 1855.

[14] "During the year 1856 some of the Magnetic Company's underground lines began
to give trouble. This was largely caused by the decay of the gutta-percha, which, laid

of the gutta-percha was comprehensibly made public later in 1857, when the British & Irish Magnetic Telegraph Company (the British & Irish), born from the merger of the two companies, published Highton's studies, which attributed much of the blame to a fungus.[15] The same phenomenon also affected the underground lines of the Electric and though the technicians consulted various studies[16] and attempted a series of remedies,[17] they came up with no conclusions

in a dry soil, lost much of its essential oil, by evaporation: its woody substance only remained – and this in a porous condition. To overcome the above difficulty Charles Bright devised an apparatus for first forcing a liquid insulating substance into the pores of the decayed gutta-percha, and then giving it an outer coating of a more solid material. The plan, however, was not found sufficiently satisfactory in practice. The authorities thereupon set themselves to consider how they could best extend their over-head system. This culminated in the absorption of the British Telegraph Company, which had exclusive rights (as before mentioned) for overhead telegraphs along the public roadways....After the above amalgamation...the underground wires were only used in places where circumstances rendered them specifically desirable", Edward and Charles Bright, *The Life Story of the late Sir Charles Tilston Bright*, Westminster, Archibald Constable and Co., n.d. pp. 72–73.

[15] "The following is the result of Mr. Highton's investigation: – With reference to my experiments on the action of the mycelium of a fungus on gutta percha, I have for some months been growing one of the class called *Agaricus campestris* in contact with gutta percha. I find as the result that the mycelium of this fungus does rapidly destroy the insulating properties of gutta percha; and in fact it appears to decompose entirely this vegetable gum....I will then communicate with the board when I have made further investigations, but at present I feel bound to say that the presence of the mycelium of a fungus, and the decaying of the gutta percha covering of the telegraphic wire, being so constantly associated together, I can come to no other conclusion than this – viz., that the mycelium of a fungus will cause decomposition in gutta percha and probably in most other vegetable productions", "Calendar of operations", *Bucks Herald*, 3 October 1857.

[16] "The reports of Mr. Edward Highton, of his examination of the underground wires of the British Telegraph Company, which had become faulty in certain spots from appar-ently local causes, were read to the meeting. From his investigations, it appeared that the decay of the gutta percha was in some cases attributable to the presence of the mycelium of a fungus, which caused its decomposition", BT, *Electric Telegraph Company*, TGA 1/7/1, *Quarterly meetings of Superintendents Reports*, 15 July 1857, p. 12.

[17] "It was considered by the Meeting that these faults were almost wholly attributable to the failure of the gutta percha joints. In most cases of failure there was evidently a want of union and the main covering of the wire, and this was considered to be chiefly due to the employment of dirty hands in making the joint. Where this want of union exists, the main covering of the wire, as it gradually dries, naturally shrinks away from the wire and sheet gutta percha at the joint. After considerable discussion, the following resolutions were adopted by the Meeting, with a view to secure more perfect joints; the propositions emanating from Mr. Varley and Mr. Culley respectively", BT, *Electric Telegraph Company*, TGA 1/7/1, *Quarterly meetings of Superintendents' Reports*, 13 January 1857, pp. 9–10.

or solutions.[18] It is true that the problem of rotting natural rubber was common to all the telegraph companies, but the Magnetic had a bigger underground network and was therefore in greater difficulty.

The Magnetic's uncertainties about the future of its own network of underground cables were flanked by its deep involvement in the preparatory stages of the first attempt to lay a transatlantic cable, which drew off much of the energy of its employees, of which Charles Bright was the clearest example.

Already in 1852, the telegraph companies led by the Magnetic had started off the race towards the western coast of Ireland.[19] While their first aim was to pick up the latest news from the States hot from the freshly moored ships, and send it rapidly in the direction of London, their long-term desire was to acquire landing rights for the transatlantic cable and a consequent exclusive for the transmission of telegrams. But on the other side of the ocean, it was only in early 1855 that work was speeded up to put Newfoundland, the point nearest Ireland, in direct connection with the rest of the American telegraph network. Meanwhile, in the summer of the same year, Cyrus Field, the entrepreneur behind this operation, set up a syndicate with John Brett to lay a transatlantic cable. While the businessmen worked on the finances, Charles Bright, together with Edward Whitehouse, a Brighton surgeon, carried out a number of experiments along the

---

[18] "The importance of the utmost care being taken in making gutta percha joints, and the trouble occasioned by those defectively made, were again pointed out to the meeting. Where gutta percha covered wires were kept permanently moist, unexposed to the effects of light and air, there appeared to exist no decay, but in dry places there was evidently a need of protection, and a coating of india rubber, or even of lead, was suggested for this purpose", BT, *Electric Telegraph Company*, TGA 1/7/1, *Quarterly Meetings of Superintendents Reports*, 15 July 1857, p. 12; "The rapid decay of Gutta Percha wire causes a heavy and constant expenditure, and although earnest and continual attention is given to the subject, our experiments have not yet led to a satisfactory method of preserving this material", BT, *Electric Telegraph Company*, TGA 2/1, *Engineer's Report*, June 1859.

[19] "Irish Channel Submarine Telegraph", *Nottinghamshire Guardian*, 3 June 1852; "Electric Telegraph between Ireland and Scotland", *Morning Post*, 20 July 1852; "Great Extension of the Magnetic Telegraph", *Morning Chronicle*, 22 July 1852; "The Anglo-Irish Submarine Telegraph", *Morning Chronicle*, 29 July 1852; "Submarine Telegraph between Holyhead and Dublin", *Hampshire Advertiser*, 25 September 1852; "Irish Submarine Telegraphs", *Dublin Evening Mail*, 20 October 1852; "The Magnetic Telegraph. Laying Down of the Cable across the North Channel", *Morning Post*, 26 May 1853; "British Electric Telegraph Company – Another Submarine Cable", *Belfast News Letter*, 5 June 1854.

Magnetic's underground system to find out if a telegraph message could be sent over a distance of more that 2,000 miles, the minimum needed for a transatlantic cable.[20] Given the positive results, the group went on to create a company for laying and managing a transatlantic cable with the support of both the financial world and the Government.[21] Finally, in November 1856 the Atlantic Telegraph Company was officially capitalized thanks to the essential contribution of the Magnetic, which also sponsored the advertising campaign in Great Britain.[22]

4.2   The Electric's new majority

As Wylde was not directly linked to the business world, he was in a position to act as a balancing factor between the two factions which had equal representation on the board. In fact, while he defended the new shareholders' interests, Wylde declared on many occasions that he was following the same line as Ricardo,[23] which is probably

---

[20] Charles Bright, *Submarine Telegraphs. Their History, Construction, and Working*, London, Crosby Lockwood and Son, 1898, pp. 23–29.

[21] "The principal features of the understanding between the government and the Atlantic Telegraph Company are, we believe, as follow: Her Majesty's vessels will assist the company as far as possible in correcting...laying down the electric cable. The government will give a fixed price of 14.000 l. per annum for the transmission of their messages until a dividend of six per cent is declared, when this sum will be reduced to 10.000 l., to be continued for 25 years. If, however, the number of government messages sent should be so large that, if charged for at the usual tariff, the amount would exceed these sums, the full price is to be paid to the company. The British government are [sic.] to have a right of prior transmission, unless the United State government should join in the contract. In that case the messages of the two governments are to be sent off in the order of their arrival at the station. The tariff of messages, when fixed and approved by the Treasury, is not to be varied during the existence of the contract", "Atlantic Telegraph Company", *London Daily News*, 24 November 1856.

[22] "A company called the Atlantic Telegraph Company has been started under the Limited Liability Act, for the purpose of carrying out the grand scheme of a submarine telegraphic Communications between Europe and America. The projectors of the undertaking are Mr. Cyrus W. Field, merchant, of New York, vice-president of the New York, Newfoundland, and London Telegraph Company, and Mr. John W. Brett, whose name is so well known in connection with submarine telegraphy in Europe; and it is intended that the company shall unite the interests of telegraphic world on both sides of the Atlantic", "The Atlantic Telegraph", *Liverpool Mercury*, 15 November 1856.

[23] "The accounts appended to this Report have been prepared more in detail than been usual in accordance with the desire expressed by the Shareholders at their last Half-yearly Meeting. ... This continued increase in the capital of the Company would, if its undertaking had, like a railway or canal, any defined limits, become a source of serious

why he had to face many attacks from the new shareholders. A first sign of the tension emerged when Robert Grimston, a Tory very close to Smith, carried out a special auditing on the final balance sheet for Wylde's first year of chairmanship.[24] A second sign of shareholder ill-humour surfaced in the meeting of 5 February 1857, when the Manchester committee headed by Chadwick handed out a circular which brought to the fore a list of problems attributed to the board: the substantial unanimity of the directors isolating voices of dissent like Chadwick's;[25] the lack of increase in profits; the lack of reduction in the expenses of maintaining/running the lines; a supposed increase in the extraordinary expenses which emerged from an engineers' report.[26] Cornered by Wylde's vigorous

---

anxiety on the part of the Shareholders; but they will bear in mind the expansive character of their system, and that it has no prescribed or ascertainable termini. So long as it can be extended with advantage to the community, and with an increase of income to the Company commensurate with the additional outlay, it may almost be said to be their duty as well as their interest to encounter the outlay for the desired extension....The net earnings applicable for dividend, consequently, gives a percentage on the larger capital expended at the rate of 7 per cent per annum, as against a percentage at the rate of 6 per cent per annum for the former period. This gradual but certain growth in the percentage of profit is highly encouraging, and affords the best assurance not only for the future maintenance of its present remunerative rate, but of its further advancement. The full value of this assurance will be best appreciated by a reference to the past experience of the Company", BT, *Electric Telegraph Company*, TGA 2/1, *Directors' Report*, December 1856.

[24] "It may be satisfactory to the Shareholders to know that a special investigation, conducted with the utmost care by the Hon. Robert Grimston, who has devoted great attention to the task, has been made into the Share Register and Transfer Departments of the Company; and, as the Directors anticipated, everything has been found perfectly regular and correct", Ibid.; "To receive M. Grimston report on his examination of the Company's books and account. Resolved that the report be received. That thanks of the directors be tendered to M. Grimston. Resolved that all future transfers be examined by a Directors", BT, *Electric and International Telegraph Company*, TGE 1/1/1, Board of Directors, 8 January 1857.

[25] Chadwick may well have been referring to the following motion advanced by the Manchester committee he represented, which was unanimously rejected by the board: "To consider the steps to be taken in consequence of the communication made to the Board at the last meeting from a committee of shareholder at Manchester. Resolved Unanimously that the request of the Committee of the Shareholders at Manchester be not acceded to", BT, *Electric and International Telegraph Company*, TGE 1/1/1, Board of Directors, 27 January 1857.

[26] "Having explained these items, the chairman went into a refutation of certain statements made in the circular published by the committee of the shareholders sitting at Manchester. He denied that there was any want of unanimity at the board, unless a general determination on the part of the directors no to allow Mr. Chadwick to have his own way in the management of the company could be considered a want of unanimity.

reply,[27] Chadwick gave expression to his vision of the company, holding that the real problem of the Electric went back to 1851, when the lines had been extended to places which gave poor returns, i.e., when Ricardo had adopted a vigorous policy of direct investment in the network.[28]

Meanwhile, a series of in and out movements had altered the network of alliances on the board. In December 1855, Morton Peto (very close to Brassey) had left,[29] and in June 1856, Stephenson,[30] whose anti-competition views were well known, came in. In October 1856, after months of absence Ricardo stepped down.[31] Nevertheless, in February 1857, following a shareholders' resolution, the board asked Ricardo to withdraw his resignation, which he did.[32] To his surprise,[33] his colleagues elected him chairman

---

It was, too, he said be unfair to state that the profits of the company had not increased, as in making that representation they suppressed the fact that their dividends had been paid without any deduction of income tax. The next charge against the directors was that they had not practised economy and reduced the working expenses of the company. He had already observed upon the nature and character of the heaviest items of their expenditure, and shown how moderate they were....Another statement in the circular was that the engineer of the company had reported that their plant was in such a bad state that an outlay of about 200.000 l. was required for further works and renewals. He denied that they had ever considered any such report", extract for the minutes of the Electric shareholders' meeting in "The Electric and International Telegraph Company", *London Daily News*, 6 February 1857.

[27] "The fact was that the Manchester committee wanted to force on the board of directors the adoption of their own views, but he – and he spoke the sentiments of his brother directors – would, with all due respect to the gentlemen from Manchester, see them in a very hot place before he would submit to any such a dictation", Ibid.

[28] "The cardinal error of the company commenced in 1851, when they stretched their wires into districts that would never pay, had it not been for that, the shareholders would now be receiving 10 per cent on their capital", Chadwick's answer in Ibid.

[29] BT, *Electric Telegraph Company*, TGA 2/1, *Balance sheets*, December 1855.

[30] Ibid., June 1856.

[31] Ibid., December 1856; "Mr. Ricardo's letter offering his resignation as Director", BT, *Electric and International Telegraph Company*, TGE 1/1/1, Board of Directors 2 October 1856.

[32] "The resolution of the half yearly general meeting inviting Mr. Ricardo to resume his seat at the Board of Directors having been brought under Consideration. Resolved that the resolution of the General Meeting be communicated to Mr. Ricardo. Mr Ricardo requested the Board to allow him a few days for consideration before returning his answer to their request. Mr. Ricardo attended the Board when the Chairman communicated to him the Resolution of the General Meeting", BT, *Electric and International Telegraph Company*, TGE 1/1/1, Board of Directors, 6 February 1857.

[33] "The Chairman, in moving the adoption of the report and statement of accounts, said he had felt anxious to retire from the chairmanship of the company, and he was rather surprised at the result of the last meeting", "Electric and International Telegraph Company", *London Standard*, 1 August 1857.

again.[34] Not by chance, his appointment came at the same time as Brassey, still an important shareholder, came onto the board, to be counterbalanced by the simultaneous admission of Grimston.[35] Back as chairman, Ricardo found himself in the same position as Wylde, having to manage a board divided into two factions. On the occasion of the June 1857 interim financial statement, Ricardo himself recalled the heightened rivalry between the shareholders and consequently the directors, and expressed the hope that with his return it could be resolved.[36] Contrary to his expectations, his second chairmanship was very brief. Following the approval of the issue of preference stock, which Ricardo had always opposed,[37] in February 1858 he stepped down definitely[38] and was significantly replaced by Stephenson as chairman, backed by Grimston as deputy-chairman. From then on the group of conservatives and small shareholders, who defended a short-term strategy at the cost of big

[34] "Mr. Ricardo attended and explained. Resolved that the number of Directors be increased by one. That Mr. Ricardo be elected a Director of the Company. That Mr. Ricardo be elected Chairman of Directors", BT, *Electric and International Telegraph Company*, TGE 1/1/1, Board of Directors, 19 February 1857.

[35] BT, *Electric Telegraph Company*, TGA 2/1, *Balance sheets*, June 1857.

[36] "What was the state of their affairs at this time? Why, the shares which were at 90 were above par, which was about 10 per cent increase on the value of their property. There had been several causes of the late depreciation of their property. One thing was that some parties in the company held too large a number of shares, but they had since been spread more, and that was removed. Another thing was the constant dissensions that had existed amongst themselves, but he hoped they had now ceased to exist", "Electric and International Telegraph Company", *London Standard*, 1 August 1857.

[37] "When I resumed the Chair, somewhere about twelve months ago, I found that you had come to this point – that a Bill was actually in print, and was proposed to be brought before Parliament, and was already announced to you as proposed to be brought before Parliament, for the creation of Preference Stock in order to obtain more capital. Now if there is any one thing more than another that I have set my face against in any enterprise in which I have been concerned, and especially over which I have exercised any control, it has been the burthening of that undertaking with Preference Stock; and I represented to my colleagues, immediately that I resumed the Chair, my impression that it was possible we might go on twelve months longer without having recourse to any such arrangements, into which they themselves had only entered with extreme reluctance and from pressure of extreme necessity", BT, *Electric Telegraph Company*, TGA 2/1, *Report of the proceedings at the Special general meeting held at the London Tavern, 8 October 1857*.

[38] "The chairman said he was compelled to resign the chairmanship of the company owing to its requiring so much of his time. He had been with them from the beginning, and regretted to leave his position, but was happy to find a successor in Mr. Robert Stephenson, who was so well acquainted with such undertakings", *Morning Chronicle*, 5 February 1858.

infrastructural investments and favoured a policy of anti-competition agreements,[39] was definitively consolidated, as is shown by the fact that Grimston took over as chairman after the death of Stephenson in 1859.[40] The same group led the company until it was wound up.

4.3   The British/Magnetic merger

In November 1856 the British and the Magnetic announced their merger.[41] At a later date Edward Bright, then secretary of the British, declared that the reason for the merger was the gutta-percha.[42] Unfortunately, here again there are no primary sources to explain why the Magnetic moved quickly to negotiate an urgent merger with the British in 1856. The terms of the agreement, extremely favourable to the Magnetic, showed that it led the negotiations in a dominant position. In fact, while the Magnetic shareholders were guaranteed a 5% dividend for the next three years, those with the British would receive 4% the first year, 4.5% the second, and only reach parity of conditions in the third year.[43] Furthermore, the newly formed British & Irish[44] would have its headquarters in Liverpool, the

---

[39] Stephenson's significant words: "In conclusion, the Directors can only assure the Proprietors that it will be their constant endeavour to continue to cultivate friendly relations with other Companies, and to study to improve the service to the public, and, by so doing, to promote the well-being of this Company", BT, *Electric Telegraph Company*, TGA 2/1, *Directors' Report*, June 1859.

[40] Grimston on Stephenson: "It is with deep regret that the Directors have to record the death of their late chairman. Mr. Robert Stephenson had long been a member of the Board, and short as was the time during which he occupied the chair, he conferred substantial benefits on the Company by his attention to their interests, and by the soundness of his advice", BT, *Electric Telegraph Company*, TGA 2/1, *Directors Report*, December 1859.

[41] "Amalgamation of Telegraph Companies", *Liverpool Mercury*, 26 November 1856.

[42] Edward and Charles Bright, *The Life Story*, pp. 72–73.

[43] "The chairman read the terms of a proposed amalgamation, agreed to and signed by the duly authorised representatives of the directors of the companies, at a meeting held on 23rd of September last, in Manchester. The first of these terms was that an act of Parliament he applied for, if necessary, to carry out the union of the two companies, to date from January, 1857; that for three years the Magnetic company's share capital shall draw, as a first charge on the net profits, a dividend of five per cent; that afterwards the British company shall draw a dividend not to exceed 4 per cent for the first year, 4 ½ per cent for the second year and 5 per cent for the third year the balance to be divided between the two; that after the three years there be an equality in the dividend; and that the chief officers and the internal management of the United company shall be at Liverpool", "Amalgamation of Telegraph Companies", *Liverpool Mercury*, 26 November 1856.

city of the Magnetic, and take over the latter's management, as confirmed by the fact that the chairman and vice-secretary were to be Ewart and Edward Bright.

After the 1855 price agreement, the telegraph companies had changed their approach to the market, putting aside their aggressive competitive spirit of the early 1850s and trying out new forms of possible integration. In 1854 the Submarine already led the way by guiding from outside the British/European merger. It was no coincidence that these two companies, linked by cross-sharing, contacted in April 1856 Cooke, who was still a director on the Electric's board.[45] And the Electric's board considered the proposal for the British/Submarine agreement serious enough to set up at once a committee formed by its most experienced telegraph men: the then chairman Wylde, Ricardo, Stephenson, Bidder and Cooke.[46] After only a fortnight a special board meeting was called for the sole purpose of giving the directors the committee's report, so that they could reach their conclusions before the next meeting.[47] On 8 May the chairman presented his personal opinion on the report, which was shared by the whole board, though it does not emerge from the minutes. The decision was taken to proceed with negotiations and a mandate to follow them closely given to a restricted group formed by the chairman, Paget, and Cooke.[48] Nothing

---

[44] "The British and Irish Magnetic Telegraph Company", *Liverpool Mercury*, 17 February 1858.

[45] "M. Cooke report on Communications which taken place with the Submarine and British Telegraph Companies", BT, *Electric and International Telegraph Company*, TGE 1/1/1, Board of Directors, 2 April 1856.

[46] "Resolved that a committee be appointed to report to the Board on the advisability of an arrangement being made with the British and Submarine Company to consist of the Chairman, Mr. Ricardo, Mr. Stephenson, Mr. Bidder and Mr. Cooke", Ibid.

[47] "To read the minute of the last Board n. 2820 as follows: that a committee be appointed to report to the Board on the advisability of an arrangement being made with the Submarine and British Telegraph Companies. Resolved that the report be received and entered on the Minutes. That the consideration of this question be postponed to the next Monthly Board Meeting", BT, *Electric and International Telegraph Company*, TGE 1/1/1, Board of Directors, 16 April 1856.

[48] "To further consider the negotiation with the Submarine and British Telegraph Companies postponed from Special Board Meeting of 16th April last. Resolved that the Board having heard the statement of the Chairman approve the same. That the Chairman be requested to proceed with the negotiations and that every effort be made by petition or otherwise effect the objects the Company have in view. That a Committee consisting of the Chairman, Lord Alfred Paget and Mr. Cooke be appointed to carry out the resolution", BT, *Electric and International Telegraph Company*, TGE 1/1/1, Board of Directors, 8 May 1856.

happened for almost two months,[49] until in July Wylde and Till were asked to draw up a report on the latest proposals received from the Submarine and the British.[50] Yet again the content of the report is unknown, but not its effects. The board decided to organize a committee together with the British and the Submarine to evaluate leasing their lines and giving their shareholders a half of the dividend payable to the Electric's partners proportionate to the agreed capital.[51] In other words, in July 1856 the Electric was seriously taking into consideration the possibility of merging with the Submarine and the British. At this point the negotiations seemed to vanish into thin air until 4 September, when the board received a letter in which the British (but not the Submarine), expressed its sorrow at not having been able to reach a satisfactory agreement for the parties involved.[52] Just at the moment in which the matter seemed over and gone, it was again brought up. After the British's communication, a letter was read from Langworthy, one of the board members, about negotiations he was following with

---

[49] In June 1856, the board received the report on the negotiations, though their reactions are not given: "to receive the report of the Committee appointed for the negotiation with the Submarine Telegraph Co.", BT, *Electric and International Telegraph Company*, TGE 1/1/1, Board of Directors, 5 June 1856.

[50] "The Chairman to report that further negotiations had been opened with the Submarine and British Telegraph Company. That General Wylde and Till be requested to prepare and lay before the Board a report upon the office of the British and Submarine Telegraph Companies and that a special Board be called on Monday 7th July", BT, *Electric and International Telegraph Company*, TGE 1/1/1, Board of Directors, 3 July 1856.

[51] "To lay before the Board the report on the offer of the Submarine and British Telegraph Companies. The minute of the Board of the 3rd July appointing the Special Meeting was read. The report submitted by the Chairman and Mr. Till was read. ... Resolved that a committee be appointed to negotiate with the Submarine and British Telegraph Companies for the leasing of the latter Companies lines with all all [sic.] advantages and engagements Charter acts of Parliament and powers upon the basis suggested at this Board namely, that a dividend not exceeding one half of the dividend payable to the proprietors of the Electric Telegraph Company be paid to the British Telegraph Company on an agreed amount of their Capital. It being understood that the object to be obtained by the above negotiation is primarily to effect an engagement with the Submarine Company for a mutual understanding with that Company either for a division of the Continental traffic or for disposing of the International Telegraph for a valuable consideration", BT, *Electric and International Telegraph Company*, TGE 1/1/1, Board of Directors, 7 July 1856.

[52] "To lay before the Board a Letter from the British Telegraph Company of the 9th August, with extract from Board minutes, expressing regret that it had not been possible to arrive at a satisfactory arrangement between the Companies", BT, *Electric and International Telegraph Company*, TGE 1/1/1, Board of Directors, 4 September 1856.

the Magnetic.[53] Evidently the directors had been sounding out via unofficial channels the margins of negotiations with the Magnetic, which had seemed to be outside the games.

Surprisingly, a month later on 2 October the directors were informed that both the Magnetic and the British intended to merge with the Electric.[54] The board appointed a committee formed by Ricardo, Chadwick and Bidder and tasked them to meet with the representatives of the two companies and choose a neutral party who could decide on equal criteria for the merger.[55] In reality, the project to create a private monopoly of British telegraphs disappeared very quickly and in early November, the board of the Electric could do nothing but take note of the already completed merger of the British and the Magnetic[56] and warn that it would be taking all necessary measures to protect its privileged position in the market.[57]

On giving the official announcement of the British/Magnetic merger, the Magnetic's chairman explained the terms of the agreement and specified that they were the result of a meeting held between the two companies on 23 September 1856. The precise date appears only in a secondary source,[58] but if correct, it means that

---

[53] "Letters from Mr. Langworthy on the negotiations with the Magnetic Company", Ibid.

[54] "Having been reported to the Board that the English and Irish Magnetic and the British Telegraph Companies have expressed a willingness to amalgamate with this Company. Resolved", BT, *Electric and International Telegraph Company*, TGE 1/1/1, Board of Directors, 2 October 1856.

[55] "To appoint a Committee with power to meet Committees of the Magnetic and British Telegraph Company to make arrangements with them for appointing an indifferent party to draw up equitable terms for the amalgamation of the three Companies, such terms to be binding on all parties. That a Committee be appointed to negotiated with those Companies on behalf of the Electric Telegraph Company with a view of carrying out the proposed amalgamation. That the Committee consist of the Chairman ex officio, Mr. Bidder and Mr. Chadwick", Ibid.

[56] "To consider the future policy of the Company, in consequence of the amalgamation of the Magnetic and British Telegraph Companies. It appearing from the report of the Chairman that the negotiations for the union of the several Telegraph Companies have terminated", BT, *Electric and International Telegraph Company*, TGE 1/1/1, Board of Directors, 6 November 1856.

[57] "That it is incumbent on this Company to take active measures for the purpose of retaining their present advantageous position and they will be prepared to consider any proposition recommended by their Chairman for the accomplishing such objects", Ibid.

[58] "Amalgamation of telegraph companies", *Liverpool Mercury*, 26 November 1856.

the terms of a possible merger had been fixed before the negotiations with the Electric over the setting up of a big private monopoly.
The Magnetic was indeed in urgent need of a merger, yet it presum-
ably rejected the Electric's favourable conditions, which were most
probably the same as those offered in July to the British. It must be
remembered that the Magnetic played with its cards close to its chest
and that the story of its ruined underground lines was not known
until 1857. Thus, its directors were able to present themselves as the
representatives of a dynamic, technologically advanced company
with a quasi monopoly over telegraphs in Ireland, just in the period
in which there was much talk about laying a transatlantic cable
between Ireland and the States.

The process leading to the merger of the two companies appears
somewhat complicated because of the expedients chosen in order to
accelerate the operation, which was approved by both the companies' shareholders at the end of November 1857.[59] It consisted in
the dissolution of the British and the Magnetic and transfer of all
their assets and liabilities (capital, contracts, debits/credits) to a new
company, the British & Irish, created for the purpose and registered with limited liability. Both the 1855 Limited Liability Act and
the 1856 Joint Stock Companies Act provided for the possibility of
starting up a joint stock company with limited liability without
needing parliamentary authorization, by simply entering the name
in a register.[60] The merger began formally on 1 January, before both
companies had finished the process of winding up.[61] This was why in
1857 the last half-yearly shareholders' meetings of the British[62] and

---

[59] Ibid.

[60] *The Limited Liability Act 1855*, 18–19 Victoria, c. cxxxiii and *Join Stock Companies Act
1856*, 19–20 Victoria, c. xlvii.

[61] BT, British Telegraph Company, TGL1/1/2, *Supplemental deed of settlement for the dissolution of the British Telegraph Company*.

[62] "The half yearly meeting of the share holders in this company was held yesterday
at the London Tavern, Bishopgate-street, the Hon. F. Cadogan in the chair. The report
contained the following statements: 'The directors of the British Telegraph Company
have the satisfaction to inform the shareholders that the amalgamation with the
Magnetic Telegraph Company is proceeding satisfactorily, and that the result, up to this
time, has been such as to meet their anticipations. The accounts of the joint company
show a surplus profit of L. 15, 312 10 s. 2 d., which after appropriating to this company's
capital the dividend to which they are entitled, and paying the amount due to the pref-
erential shareholders of the company, will leave a dividend at the rate of 4 per cent, per
annum upon the original capital of the British Telegraph Company. The arrangements
for the complete exchange of the shares are going forward, and the exchange will be
effected at the earliest moment, of which due notice will be given to each proprietor.

the Magnetic[63] were held,[64] where the last formalities were approved to allow the final passage to the new company,[65] sanctified with the first meeting of the shareholders of the British & Irish in February 1858.[66]

The directors trust that the result of the first half year's working of the amalgamated company will show, that the advantages of union, by increasing receipts and diminution of expenditure, has not been miscalculated'. On the motion of the Chairman, the report was unanimously adopted. The Chairman also moved a resolution (which was also adopted), declaring a dividend upon the original shares of the British Telegraph Company, at the rate of 4 per cent. The meeting then separated", "The British Telegraph Company Meeting", *Morning Chronicle*, 1 September 1857.

[63] "English and Irish Magnetic Telegraph Company", *Liverpool Daily Post*, 29 August 1857.

[64] Ibid.

[65] "Yesterday, a meeting of the proprietors of the English and Irish Magnetic Telegraph Company (limited) was held at the Clarendon-rooms, for the purpose of passing a resolution winding up voluntarily the affairs of the company, under the provisions of the Joint Stock Companies Act 1856 and 1857....The Chairman said the directors felt sorry to be obliged to call the proprietors together, and to give them the trouble of meeting, merely to have submitted to them a resolution which they had already passed. At the time of the previous meeting, it was under an act of parliament which was found to be inadequate or the purposes for which it was designed. And not, however, remedying the defects had been passed during the present session, and it therefore became necessary, in consequence of that supplementary act, that this meeting should take place and that the resolution should be again submitted. The resolution would be read and as the meeting was special, no other business could be transacted", Ibid.; "An extraordinary general meeting of the proprietors of the English and Irish Magnetic Telegraph Company (limited) was held yesterday at the Clarendon Rooms, South John-street, for the purpose of confirming a special resolution passed at the extraordinary general meeting of the said company, held on the 28th day of August last – That the said company be wound up voluntarily, under the provision of the 'Joint Stock Companies Act, 1856–1857', and for the purpose, in case such resolution should be confirmed, of appointing a liquidator, or liquidators, to wind up the affairs of the said company; and of adopting such other measures as might be necessary for carrying the said special resolution into effect. This course was necessary to enable the property of the company to be legally transferred to the British and Irish Magnetic Telegraph Company, under the united system of management, but which was formerly worked by separate companies", "The English and Irish Magnetic Telegraph Company", *Liverpool Mercury*, 6 November 1857.

[66] "The first ordinary general meeting of this company was held yesterday at the Clarendon Rooms, South John-street, for the purpose of receiving a statement of the accounts, electing directors, and transacting other Business....The Chairman, after stating the object of the meeting called on the secretary, Mr. E. B. Bright, to read the report, which was as follows: 'During the past year the arrangements for the complete union of the lines of the English and Irish Magnetic Telegraph Company and the British Telegraph Company, sanctioned unanimously by the meetings of the shareholders of each company in November, 1856, have been fully carried out, and the results, in the opinion of the directors completely support the anticipations formed when the union was first negotiated'", *The British and Irish Magnetic Telegraph Company*, *Liverpool Mercury*, 17 February 1858.

The official reason for the merger had been for both companies the need to reduce running expenses at once.[67] The integration of the two networks would allow offices and double lines to be closed and redundant personnel dismissed.[68] An operation over the short period that would please the small and medium-sized shareholders typical of both companies by allowing an increase in the dividends.[69] With the same aim of pleasing the shareholders, the British & Irish, though avoiding further corporate organization, favoured the birth of a telegraphic centre, which under its sponsorship included both the Submarine and the London District Telegraph Company (the District), a company created in 1859 with the backing of numerous shareholders and directors of the British & Irish and the Submarine.[70] Though the Submarine was formally independent,

---

[67] "The Chairman replied that ... would be a large saving of expense, probably amounting to £ 12.000. The lines of both companies were running at present side by side through many districts; they would also obtain a connection with the continent, which they had not at present", "Amalgamation of Telegraph Companies", *Liverpool Mercury*, 26 November 1856; "He [the Chairman] might mention, also, as a satisfactory piece of information, that by the two companies working together a saving had been effected in the six months of £ 7,500", "English and Irish Magnetic Telegraph Company", *Liverpool Daily Post*, 29 August 1857.

[68] "A saving in working expenses has been effected (the item of maintenance of works only excepted) to an extent exceeding £7,000 when compared with those of the two companies prior to the lines being joined together. Further reductions are still in progress, some necessarily requiring to be gradually carried out. These results of the last year's working are mainly to be referred to the greater advantages and facilities offered to the public by the increased efficiency of the united system, and to the discontinuance of 16 duplicate stations, and in some instances the double staffs employed, in towns where the Magnetic and British Telegraph Companies formerly competed with one another", "The British and Irish Magnetic Telegraph Company", *Liverpool Mercury*, 17 February 1858.

[69] "A saving in working expenses has been effected (the item of maintenance of works only excepted) to an extent exceeding £7000 when compared with those of the two companies prior to the lines being joined together. Further reductions are still in progress, some necessarily requiring to be gradually carried out. These results of the last year's working are mainly to be referred to the greater advantages and facilities offered to the public by the increased efficiency of the united system, and to the discontinuance of 16 duplicate stations, and in some instances the double staffs employed, in towns where the Magnetic and British Telegraph Companies formerly competed with one another", "The British and Irish Magnetic Telegraph Company", *Liverpool Mercury*, 17 February 1858.

[70] The chairman of the District was Samuel Gurney, who was also a director and shareholder of the Submarine. Other board members were Charles Kemp Dyer, shareholder and director of the British and Irish, and one of the first telegraph technicians, and John Watkins Brett, founder of the Submarine, pioneer in submarine telegraphy, and director of the Mediterranean Extension Telegraph Company. The board also included

it was linked to the British & Irish both by cross-holdings[71] and a contract allowing the latter exclusive passage along its cables to the European continent.[72] The District instead would have to deal exclusively with transmitting telegrams in the capital, linking up with stations not more than four miles apart. An innovative urban service offered London businessmen two instruments: urban telegrams for numerous ordinary everyday administrative duties and extra-urban telegrams which could be sent from any of the District's numerous station offices in the British & Irish's network.[73] An agreement between the two allowed the District synergies with the network,[74] which had its telegraph centre in a new main seat in

political figure linked to the city: Robert Taylor, vice-chairman and member of the Metropolitan Board of Works, William Anderson Rose, London businessman and Conservative mayor of London from 1862. "An intimation has been published on the Stock Exchange that the District Telegraph Company will commence operations on 1st February", *Cheltenham Chronicle*, 17 January 1860.

[71] At the moment of the merger, a substantial packet of British shares was still held by three big shareholders of the Submarine.

[72] "The Communications with the continent is effected by mean of the cables of the Submarine Telegraph Company, the wire of which are exclusively connected to this company's wires at Dover", *The British and Irish Magnetic Telegraph Company, Liverpool Mercury*, 17 February 1858.

[73] "The London District Telegraph Company, with some hundreds of offices, is destined to play a very important part in the communication of the metropolis; as for we believe the small sum of 4d. a tradesman in Regent-street will be able at once to order any description of goods from a wholesale house in Wood-street, Gresham-street, or the surrounding districts, and receive them with the least possible delay; or a person detained in the City by unforeseen business can communicate to his family not to expect him at home at the usual hour; or, indeed, the cases which may arise are so numberless that we need not speculate upon them", "Submarine Telegraphs", *Morning Post*, 11 July 1859.

[74] "The directors of the London District Telegraph Company (limited) have issued their report preparatory to the meeting on the 23rd August, which details the progress of operations since the formation of the enterprise. Of the 12.000 shares of 5l. each authorised to be raised, the deposit of 1l. per share was paid upon 10.740. This sum it is stated is more than sufficient to discharge all the liabilities upon contracts which the company have entered into, as well as the preliminary and current expenses, and the directors have determined that no call shall be made upon the shareholders until absolutely required by the progress of the works, which will be vigorously carried on. The entire construction of the works will be completed within the original estimates, and an agreement has been concluded with the British and Irish Magnetic Telegraph Company for an interchange of traffic, by which a large amount of business will be secured. Arrangements have also been made with one of the dock companies for the receipt and transmission of messages; and a contract has been completed with the Surrey Canal and Dock Company for the exclusive right, for a term of years, to erect poles and wires on the pathway of their canal. Eleven district stations are at first to be opened, and many of these have been secured in prominent positions; the whole to

Threadneedle Street London,[75] where the central offices of the
Submarine and District were also located. The building was sited
strategically at the heart of the commercial and business area of the
city, not far from Founder's Hall, the historic seat of the Electric. It
was organized for dealing with telegraph messages, with the offices
of the three companies perfectly integrated by means of a series of
internal goods hoists which allowed an easy exchange of telegrams.[76]
Significantly, in the same period the Electric left its headquarters
in Lothbury for a more spacious and welcoming building custom-
built by Stephenson and Hunt in Great Bell Alley,[77] now known as
Telegraph Street. The new central station also answered the specific
needs of the telegraph service. Gas lighting and big windows to allow
a good visibility both day and night, spacious rooms with high ceil-
ings to help ventilation and a telegraph gallery reserved for female
workers, who had to be separated from their male colleagues.[78]

---

be ready for opening on the 1st of next January. It appears from the statement in the
report that young women are to be employed as telegraphists, under the instruction of
an efficient matron. According to the balance-sheet 9.000l is placed out at deposit, and
the balance at the bank is 234l.; making a total of 9234l", *London Standard*, 20 August
1859.

[75] "The directors have been for some time anxious to secure a permanent site for a
central station in London, and have succeeded in obtaining, with this view, the Baltic
Coffee House, with adjacent premises, facing the Royal Exchange. ... When arrangements
for opening these premises have been fully completed, the company will be enabled
to give up three of their present London stations in the neighbourhood, namely, Old
Broad-street, Cornhill and Throgmorton street, by dispensing with which a considerable
saving will accrue, and the business of the company in the city be much consolidated,
while at he same time better office accommodation will be obtained", "The British and
Irish Magnetic Telegraph Company", *Liverpool Mercury*, 17 February 1858.

[76] "The Submarine Company are removing their offices from Cornhill to most extensive
premises, built entirely for telegraphic purposes, on the site of the old North and South
America Coffee-house, Thread needle-street, and which will be occupied conjointly
by that company, the British and Irish Magnetic Telegraph Company, and perhaps others: thus
forming one grand centre for telegraphic Communications. The offices of the various
companies are very commodious, and are all connected one with the other with lifts
from the message office to the various instrument-rooms, from whence communications can at once be made between all parts of England, Scotland, and Ireland, and the
Continent of Europe", "Submarine Telegraphs", *Morning Post*, 11 July 1859.

[77] "For the purpose of effectively carrying on the business of the Company, it had
become necessary to erect new offices in Bell Alley and at Camden Town, both of which
are now far advanced towards completion", BT, *Electric Telegraph Company*, TGA 2/1,
Directors' Report, June 1858.

[78] "There has been just completed for the above company a large and substantial
building, from the designs of Messrs. Hunt and Stephenson, of Parliament street. The
most interesting portion of this new building is the Telegraph Gallery, of which we
give an Engraving. Messrs Hunt and Stephenson have had to apply architecture to the

At the beginning of 1860, these two buildings were the symbols of the new duopoly of the telegraph market.

4.4   The consolidation of the duopoly

In 1857 the telegraph market was shared between two main parties, the Electric and the British & Irish group, which gave no signs of open competition, but on the contrary set up many forms of cooperation. For example, on the occasion of the opening of the first transatlantic cable, the two made an agreement to share in equal parts the telegraph traffic to and from the States,[79] though unlike

---

novel requirements of the telegraph, and have, for the sake, principally, or obtaining light, extended this great telegraphic gallery over the whole top of the building. It is well known that the cause of female labour owes much to this company. The directors have developed a new branch of female employment, and one which appears admirably suited to their capabilities and comfort. The foreign gallery is worked by male telegraphists, nearly all foreigners; but the great gallery, in which the telegraphic business of the United Kingdom is performed, is worked solely by young females. There are, at the present time, ninety-six or ninety seven young ladies engaged daily; and, apart from the telegraphic requirements in the gallery, every arrangement appears to be made for their comfort and privacy. It may be interesting to give the dimensions of this unequalled telegraphic gallery: – the room is about 80 feet in length, 30 feet in width, and 30 feet in height. Further details of this arrangement we refer to our Engraving. It is lit from the roof with a steady northern light, and by large windows at the sides and ends: these serve also for ventilation. Two large sun-burners are provided, and a gaslight, with shade to each instrument", "Electric and International Telegraph Company, Great Bell Alley", *Illustrated Times*, 14 April 1860.

[79] "The Chairman reported he had entered into an agreement with the Chairman of the British and Irish Magnetic Telegraph Company for the conveyance of messages in Great Britain passing by the prospected Atlantic Telegraph to and from the United States as follows and read correspondence. Enter agreements. Resolved that the agreement is approved and confirmed. That a copy of the memo of agreement be sent to Mr. Brown the Chairman of the Atlantic Telegraph Co. and he be informed it has been confirmed by the Board", BT, *Electric and International Telegraph Company*, TGE 1/1/1, Board of Directors, 1 July 1857; "The Chairman read the correspondence with Mr. Brown and reported he had entered with an agreement with the Chairman of the British and Irish Magnetic Telegraph Company for the Conveyance of messages in Great Britain passing to and from the United States by the Atlantic Telegraph as follows. Resolved that the agreement is approved and confirmed. That a copy of the Memo of agreement be sent to Mr. Brown the Chairman of the Atlantic Telegraph Co. and he be informed it has been confirmed by the board", BT, *Electric and International Telegraph Company*, TGE 1/1/1, Board of Directors, 21 July 1857; "To consider the agreement with the Magnetic Telegraph Company for the conveyance of Atlantic Telegraph Company's messages. Resolved. That while the Board acknowledge the propriety of adhering to the terms of the agreement entered into on 10 July 1857 it is considered advisable to await a reply to the Chairman Letters of the date 28th May last, with the view of entering into a more comprehensive arrangement", BT, *Electric and International Telegraph Company*, TGE 1/1/1, Board of Directors, 2 June 1858.

the Magnetic/British & Irish, the Electric wanted to avoid any financial involvement.[80] In addition, in order to cut down on running
costs, both companies chose to share offices in less prestigious loca-
tions, for example, inside railways stations,[81] and encourage the
use of each other's lines in case of need.[82] Finally, after the 1855
cartel, the price of telegrams was kept more or less stable by means
of further, more specific agreements.[83] Sharing a common policy in
favour of distributing high dividends, the two companies de facto
were running an informal cartel. During the late 1850s, the Electric
followed a shrewd policy of containing costs and increasing income,
thus chalking up a more or less constant increase in overall profits.
They were, however, widely distributed among the shareholders,
who between 1856 and 1859 pocketed dividends between 6%

---

[80] "The directors have watched with much interest the measures now being taken to
establish a telegraphic communication with the United States, and although they have
not considered themselves justified in investing any portion of the property of the
shareholders in the enterprise of the Atlantic Telegraph Company, they nevertheless
anxiously desire the success of an undertaking calculated to confer such immense
benefit on the two countries, to be, by its means, so closely united, and the successful
accomplishment of which will, doubtless, give a vast impetus to the telegraphic system
generally. An arrangements has been entered into with the Atlantic Telegraph Company,
in conjunction with the Magnetic Telegraph, on terms mutually beneficial, which will
ensure a cordial cooperation and render the three systems available for providing the
requisite facilities for the transmission and reception of the American correspondence",
BT, *Electric Telegraph Company*, TGA 2/1, *Directors' Report*, June 1857.

[81] "It appeared from the evidence of the Superintendents that at stations where both
Companies are in the same office, this Company obtains the majority of the messages",
BT, *Electric Telegraph Company*, TGA 1/7/1, *Quarterly meetings of Superintendents' Reports*,
17 June 1859, p. 5.

[82] "The deputy Chairman brought to the attention of the Board the laying down
of the Submarine line from Dover to Emden and the steps now being taken by the
Magnetic Tel Co for extending the line to London. Resolved that it be referred to the
Deputy Chairman to make such offer to the British and Irish magnetic Telegraph
Company for carrying them messages between London and Norwich as he may done
fit", BT, *Electric and International Telegraph Company*, TGE 1/1/1, Board of Directors, 3
November 1858.

[83] "Alteration of Tariff to Ireland. From (?) to all parts of Ireland to include the Magnetic
Company Charge to (?) to Dublin only. Chairman reported he had made arrangements
for altering the tariff to Ireland with Chairman of the Magnetic Company", BT, *Electric
and International Telegraph Company*, TGE 1/1/1, Board of Directors, 27 January 1858;
"Report on arrangement for uniformity of charges for Intelligence, settled with the
Magnetic Telegraph Co.", BT, *Electric and International Telegraph Company*, TGE 1/1/1,
Board of Directors, 5 May 1858; "It was reported that the Intelligence arrangements
with the Magnetic Telegraph Company has been settled and adopted", BT, *Electric and
International Telegraph Company*, TGE 1/1/1, Board of Directors, 7 July 1858.

and 8%.[84] Only a marginal part of the profits was invested in repairs, while no investments at all were made in expanding the network. What had been put in the war chest since 1858 was for a reserve fund to be used for repairing breakdowns in the company's future submarine cables.[85]

Unfortunately, the same data is not available for the British & Irish, but the directors' annual reports, and especially the fact that between 1856 and 1859 the company distributed a dividend between 4% and 5%, according to the merger agreements,[86] lead us to think that

---

[84] More precisely: 6% in June 1856, 7% in December 1856, 8% in June 1857, 8% in December 1857, 7% in June 1858, 6% in December 1858, 6.1% in June 1859, 3.5% in December 1859, 3.5% in June 1860 and 3.5% in December 1860. The sharp drop in the dividends between 1859 and 1860 was not due to lower profits, but to the creation of a trust fund for covering the costs of maintaining and replacing submarine cables, particularly between Great Britain and Holland. BT, *Electric Telegraph Company*, TGA 2/1, *Balance Sheet* from June 1856 to December 1860.

[85] "The efficient maintenance and stability of the land wires and poles have been carefully attended to by the Directors and their officers. Superior insulation, renewals, where necessary, and considerable general improvements have been duly provided for out of the current expenditure, and the Directors have confidence in stating that at no previous period in the history of the undertaking has the service been so satisfactorily performed; but there is one portion of the system, viz., the Submarine Cables, which continues to give the Board much anxiety, and requires work, the rapid deterioration cannot be guarded against, or met as it arises from day to day, as on land. Casualties of an unexpected and baffling nature happened, and must be expected to recur more frequently as the cables become older, and it is evident that the time will eventually arrive when an entire renewal must be met out of revenue. To provide for this inevitable future contingency, and to prevent an undue stress on the profits of any particular year for an apparent immediate advantage, the Directors recommend a steady continuance of the sound and judicious policy which has already received the deliberate sanction of the Shareholders, a policy which the enlarged experience of the Directors of the delicate nature and working of Submarine Cables fully confirms. The Board therefore recommends that the entire balance of the half year's profits after providing for the dividend, and amounting as stated above to £ 6.690 18s. 9d., be added to the Trust Fund, which was established for that purpose in 1858, and which already amounts, with interest, to £ 7.539 12s. 10d. This fund has been duly invested, and is available for a sudden emergency or renewal of cables, as circumstances may require", BT, *Electric Telegraph Company*, TGA 2/1, *Directors' Report*, June 1859.

[86] "At a meeting of the directors of the British and Irish Magnetic Telegraph Company (Limited) held yesterday in Liverpool, a dividend was declared at the rate of 5 ½ per cent, on the Magnetic section share Class A, and of 4 ½ per cent on the British section of shares Class C, a large reserve being carried over. The receipts were stated at 71,744 l 14s 11d, and the expenses at 39,426 l 9s 1d", *London Daily News*, 8 February 1858; "The balance available for dividend was £ 12.600 (which is about £ 1000 per annum on shares of Class A, and 4 per cent per annum on those of Class C)". "British and Irish Magnetic Telegraph Company", *Liverpool Mercury*, 16 February 1859; "This increase [of expenses] has occasioned an excess over the stipulated amount for maintenance agreed

Table 4.1   Accounting data for the *Electric* 1855–1860 (in £, with decimals)

	Expenses	Receipts	Balance	Dividend	Fund	Expenditure on renewals
June 1855	46,269	67,689	21,420	18,282		
December 1855	54,860	77,239	22,379	21,501		
June 1856	53,607	78,516	24,909	22,632		
December 1856	56,756	84,737	27,981	24,754	2,600	
June 1857	56,129	86,877	30,748	28,291		3,319
December 1857	64,535	93,857	29,322	28,583		
June 1858	56,388	88,059	31,671	25,670	6,100	
December 1858	64,932	89,579	24,647	23,396	1,140	
June 1859	65,489	97,916	32,427	25,735	6,690	
December 1859	67,364	103,758	36,394	27,715	8,678	
June 1860	66,397	100,991	34,594	28,975	5,618	
December 1860	71,808	106,925	35,117	28,975	6,362	

Source: BT, *Electric Telegraph Company*, TGA 2/1, *Balance Sheets* from June 1855 to December 1860.

Table 4.2   Development of the miles of lines and telegraph wires

	Electric miles of lines	Electric miles of wires	British and Irish miles of lines	British and Irish miles of wires	Total miles of lines	Total of wires
1856	5,398	28,627				
1857	5,637	29,498	3,248	15,565	8,885	45,063
1858	6,103	30,733	3,441	15,888	9,544	46,621
1859	6,272	31,346	3,656	16,732	9,928	48,078
1860	6,541	31,148	4,223	n.f.	10,764	n.f.

Source: BT, Electric Telegraph Company, TGA 2/1, Engineer's report from June 1856 to December 1860; The British & Irish Magnetic Telegraph Company, *Liverpool Mercury*, 17 February 1858; British and Irish Magnetic Telegraph Company, *Liverpool Mercury*, 16 February 1859; British and Irish Magnetic Telegraph Company, *Liverpool Mercury*, 25 February 1860.

like the Electric it adopted a policy of scarce reinvestment of profits in network extension. The result was that both companies slowed right down in line building. While there were more than 8,000 miles of lines in 1855, five years later only an extra 2,000 or so had been added. The following table shows that both companies were equally responsible for the down-curve in expansion.

Another consequence of the strategy was that the companies frequently drew from the running costs to finance both ordinary and extraordinary line maintenance. Only tapping their profits occasionally, in cases of urgent and serious repairs to infrastructures. The Electric had by then many old lines, where the rotting wood and cracked ceramics were constantly being replaced,[87] though repairs were carried out gradually and could be spread over various balance sheets.[88] In the case of the British & Irish, profits had to be used to replace the network of underground cables which had collapsed between in 1859 and 1860,[89] as had been foreseen by the Bright Brothers.

---

to on the amalgamation, which is equally borne by both classes of shareholders, and causes a deduction of one-half per cent for the half year ending Dec 31, 1859, from the five per cent dividend to which the A shares would have otherwise been entitled, which is equivalent to one-quarter per cent, for the year", "British and Irish Magnetic Telegraph Company", *Liverpool Mercury*, 25 February 1860.

[87] "The renewal of decayed timber, which forms an important element in the cost of maintenance, has now assumed a tolerably constant average, and the sums at my disposal will enable me before winter to place the whole of you lines in a state of efficient repair. The improved insulation which has been recently adopted continues to give the highest satisfaction, and is now in general use throughout the Kingdom, especially on long circuits", BT, *Electric Telegraph Company*, TGA 2/1, *Engineer's report*, June 1858; "Your lines throughout the country are maintained generally in the same condition as has been described in former reports; a systematic inspection of the lines is made annually, and all necessary repairs and renewals are unstintingly executed during the summer and autumn months. The expenditure under this head is large, and is chiefly attributable to the decay of two perishable materials, timber and gutta percha. Great attention has been given to their preservation, and for many years past experiments have been carried out on an extensive scale with a view to their more economical maintenance. The lines wires, instruments and insulation are singularly imperishable, and experiments show that by the use of iron it is possible to make the poles equally so; but the saving in maintenance is scarcely sufficient to compensate for the increased interest on capital", BT, *Electric Telegraph Company*, TGA 2/1, *Engineer's report*, June 1860; "The cost of maintenance and renewal have been very great this year; and as such works are best prosecuted in the summer season, the larger portion of the expenditure falls on the latter half of the year. The increase, independently of that attributable to additional mileage, is principally occasioned by the extensive decay of gutta percha and timber. I endeavour, as far as possible, to dispense with the use of the former material, or, where employed, to protect it by asphalt. The timber employed in 1851 and 1852, when very extensive additions were made to the system, was native larch, and this in its unprepared state has not proved durable. The simultaneous renewals of a great deal of this work, together with that of some the older lines, in which foreign timber was employed, has pressed very heavily on the maintenance. During the year one eight part of the whole of your lines has been overhauled and repaired by gangs of men set apart for that operation", BT, *Electric Telegraph Company*, TGA 2/1, *Engineer's Report*, December 1860.

[88] The only exception was the renewal of the lines in the first half of 1857, where, as Table 4.2 shows, a special fund had to be tapped.

[89] "In presenting to the shareholders the accounts of the company for the past year the directors have to report that in consequence of the sudden decay of the gutta percha

The lack of a long-term investment strategy determined not only a stop to the expansion of the landline network but also a considerable technological stagnation. In the late 1850s neither of the two companies was able to introduce any technological innovations capable of displacing competitors and helping raise revenues or lower costs, like the Henley-Foster magnetic telegraph in the early 1850s. True, the Electric did continue to try out continual improvements on the form, techniques and materials for insulation,[90] the size and

---

with which the underground wires of the company's main commercial lines were insulated, they have found it necessary to take them up and substitute the system of wires on poles. During this change your directors have been subjected to great anxiety, owing to the difficulty experienced in maintaining communication by the underground wires while the others were being erected. Notwithstanding every effort, a large portion of that traffic requiring most expedition was temporarily lost; but during the last three months, since the improvements have been carried out, your directors have much pleasure in stating that the traffic referred to has been recovered. The cost of the alteration has been more than paid for by the proceeds of the gutta-percha covered copper wire, and has not therefore constituted any charge against the company further than the increased outlay in maintaining the lines in working order during the change", "British and Irish Magnetic Telegraph Company", *Liverpool Mercury*, 25 February 1860.

[90] "Mr Ansell introduced a new insulator, a modification of Brett's and the Prussian plan, which was not considered to have advantages over the previous forms. A discussion arose on the advantages of suspended and erect insulators. The general opinion appeared to be in favour of the erect or inverted plan on new lines, although it was considered inadvisable to use them on existing lines....Mr. Culley reported that the breakage of glass, though greatly diminished, still continued, while the state of the earthenware remained very satisfactory", BT, *Electric Telegraph Company*, TGA 1/7/1, *Quarterly meetings of Superintendents' Reports*, 9 January 1856, p. 10; "Mr. Clark introduced a new form of insulator, in accordance with a promise made at a previous meeting. It was in white porcelain, in form resembling the Prussian insulator, or the still older insulator of Brett and Little, used on the Whitehaven lines. A full discussion took place on the several points of novelty it presented, especially its small point of bearing for the line wire, the introduction of thin sharp edges to enable the moisture to disappear quickly, the addition of an internal ringed cylinder of small diameter, and the use of glazed iron as a support, and on its general proportions. Upon the whole it was very highly approved", Ibid., 10 April 1856, p. 15; "As it was proposed to re-insulate important circuits with porcelain inverts to some considerable extent, a large quantity of old stores would necessarily become available; and it was explained that these would all be returned to York Street stores for the purpose of being thoroughly tested. Faulty stores would be rejected, and those in a sound condition would be re-issued for use on short and unimportant circuits", Ibid., 13 January 1857, p. 12; "Mr. Clark stated that since the last meeting very great improvements had been made in the material of which the Inverts were manufactured. The earlier supplies had been found, on subjection to a very severe test, to be slightly defective. It is true that the test employed was most severe; yet, as samples had been found which remained perfect under it, it had been resolved, at all risks, to compel the manufacturers to produce Insulators which should be completely non-absorbent. Some thousands of Inverts, slightly faulty as above, had accordingly been rejected and thrown on the hands of the makers, and the consequence

conservation of the poles[91] and line installation.[92] Though all this research and the ensuing debates led to progressive improvements in line insulation, they did not bring about any drastic technological changes.[93] As for telegraph apparatus, the superintendents' long debate about the opportunity of replacing the two-needle telegraph introduced in the 1840s with a more rapid and efficacious model ended with the adoption of the Morse printing telegraph, thus avoiding a more delicate automatic machine with far superior transmission potential.[94] The British & Irish not

was, that we were now receiving, from two houses, an article *of the most perfect description*, specimens of which were submitted. It was reported to the meeting that White Porcelain Corrugated *Cones* for use at Bridges had been prepared; also, White Porcelain *Terminal* Insulators of the ordinary make. At Girder Bridges the ordinary Invert could of course be employed", Ibid., 15 July 1857, p. 10.

[91] "The meeting discussed the process of preserving timber introduced by Dr. Boucherie, and also the present system of repairing poles, and regretted the circumstances which necessitated the use of green larch for repairs. ... the great economy of substituting tar for paint as a coating for poles was shown. Superintendents were requested to adopt its use as much as possible", BT, *Electric Telegraph Company*, TGA 1/7/1, *Quarterly Meetings of Superintendents Reports*, 11 July 1856, pp. 13–14.

[92] "The attention of Superintendents was directed to this subject, an accident, attended with fatal results, having occurred from the poles being too near the line. In this case the man was caught by a pole whilst leaning over the side of the engine. The poles should in all cases be fixed at a distance of not less than four feet from the rails", Ibid., 13 January 1857, p. 15.

[93] BT, *Electric Telegraph Company*, TGA 2/1, *Engineer's Report* from June 1856 to December 1860.

[94] "Mr. Clark called the attention of the Meeting to the fact that the gradual introduction of the Morse system would naturally throw many double needle instruments out of use. He was therefore anxious to avoid purchasing more new double needles at present, but whished to reduce the spare stock as much as the exigencies of the service would permit. A circular will shortly be sent to Superintendents, requesting them to name those stations at which it is desirable that one or more spare instruments should be placed in order to meet cases of emergency. These lists will be considered both separately and comparatively, and the location of such spare instruments as may be found necessary decided on. The spare stock will then be limited to these stations and to the depots. The stock at the depots must also be brought down to some nominal standard, and all extra spare instruments will then be called in", BT, *Electric Telegraph Company*, TGA 1/7/1, *Quarterly Meetings of Superintendents Reports*, 13 January 1857, pp. 13–14; "a long discussion followed, on the relative merits of the two machines [Bain and Morse apparatus], and, as at the last meeting, there was much difference of opinion. Messrs. Ansell, Preece, and Varley reported on the success attending the working of Morse machines. Mr. Newman, on the contrary, considered them more troublesome and more difficult to work than Bain's instruments. He had only had two Morse machines, which were with difficulty worked for some time, but which were ultimately replaced by Bain's machines. A report of Mr. Fuller's, in favour of the latter instrument, was also read to the meeting", Ibid., October 1857, p. 16; "the extensive introduction of the printing instrument is also adding very greatly to the Company's power of rapid transmission;

only introduced no noteworthy innovation but had to reconvert its entire underground cable network, on the cutting edge in the early 1850s, now needing to be replaced by something safer and more economical.

Where both landline companies showed greater technological progress, backed by the only extraordinary investments of the period, was in the laying and running of submarine cables. In this activity, which in theory was to be marginal, both the Electric and the Submarine moved on a big scale, side by side with the emerging giants. The Electric, frustrated by the continual breakdowns of its lightweight cables across the English Channels caused by the passage of merchant ships, decided to pass over to laying heavy cables. The light ones were low cost because they were made of a sole internal conductor with a very light cover. The heavy ones contained more conductors within an iron wire sheaf and were far more expensive, but at the same time more resistant to possible mechanical agents from outside.[95] In other words the heavy cables

---

and large as the increase of business has been, the Board are enable to state that at no previous period were they more prepared for a considerable addition to it, and they will endeavour by constantly availing themselves of every improvement which science and experience can suggest, and by carefully studying the public interests, to improve the character for regularity and despatch which the company has already obtained, and by which alone it can continue to prosper", BT, *Electric Telegraph Company*, TGA 2/1, *Directors' Report*, June 1860.

[95] "Your submarine Communications with the Continent continue to work satisfactorily; but that with Ireland was entirely suspended for more than three weeks by the failure of both cables. At the present time two cables are working well to the Continent, and one to Ireland. Two Continental cables and one Irish cable are under repair. The condition of the whole, in point of insulation, is admirable; the gutta percha appears likely to remain unaffected by time, the only decay that manifests itself being in the external iron coating. Although in the first instance, and in the absence of experience, it was thought prudent to obtain security from interruption by adoption of separate single cables, the experience we have now obtained of the strains and accidents to which submarine cables are liable, renders it necessary for the Directors immediately to consider the propriety of strengthening the present cables, either singly or by laying them together, or by the adoption of a new and larger cable and the employment of these elsewhere. Thereby reducing the future annual expenditure", BT, *Electric Telegraph Company*, TGA 2/1, *Engineer's Report*, December 1857; "The expense attendant on the maintenance of light Cables has been found so onerous, and the constant liability to interruption from Fishing Vessels and Anchorage has proved so annoying, that I have, in designing a new Cable, recommended that it should be of a description larger and stronger than heretofore manufactured. Its weight being nearly ten tons per mile", BT, *Electric Telegraph Company*, TGA 2/1, *Engineer's Report*, June 1858; "The large Dunwich and Zandvoort Cable, one of the smaller Cables, and the Holyhead and Dublin Cable are working well. The Zandvoort Cable was broken by anchorage on the 11th July, but

were a long-term investment, signalling a clear counter strategy. The Submarine, which had already in 1852 laid a heavy cable in the Channel with four conductors, went on investing in brief links between Great Britain and nearby countries. Between 1856 and 1859 it placed a six-wire cable between Dover and Ostend, a two-wire one between Cromer and Emden in the Duchy of Hanover, and a cable with three conductors between Cromer and Helgoland, in Denmark.[96] The sector of submarine telegraphs was expanding by leaps and bounds, catching and holding the interest of the press, politicians and businessmen. Logical, therefore, that the two great telegraph groups concentrated their scarce investments in the sector that was really pulling ahead.

However, international traffic with its links to submarine cables, was only of concern to a minority share of telegraph users. Given the high prices and scarce development of the landlines in less productive areas, plus the technological stagnation, all factors reinforced by the duopoly, telegraph users grew more and more dissatisfied. And they, mainly businessmen, began to clamour for lower prices and a greater propagation of the service.

---

restored on the 13th of the same month; it was again wilfully damaged by an axe on the 22nd July, but it was repaired in four days. One of the wires was again injured in October, but repaired in two days, without any interference with the communication on the other wires. All the above injuries are near the English coast. The smaller Cable has been repeatedly broken and repaired. At the present time it works as efficiently as when laid down in 1853. The communication with Ireland was restored in November last", BT, *Electric Telegraph Company*, TGA 2/1, *Engineer's Report*, December 1860.

[96] "As we write the contractors of the Submarine Telegraph Company are engaged in laying down a new line of telegraph from Cromer, in Norfolk, to Helgoland, for the accommodation of Denmark and the north of Europe – the cable having, after being fully tested, been shipped on Thursday last from the yard of the manufacturers, Messrs. Glass, Elliot and Co., at East Greenwich. Although the company has only been established a very few years, it will, when the line now laying down is completed, or say within a week, have five lines of telegraph in connection with the Continent at work. The first cable by which a submarine communication was effected with the Continent was laid down between Dover and Calais, and contained four wires; and shortly afterwards, in order to meet the requirements of the Central Germany, which demanded a more direct communication, another cable containing six wires was laid from Dover to Ostend. .... For the advantage of a telegraph and for the further accommodation of the traffic of North and Central Germany a cable with two wires was laid down between Cromer and Emder, in Hannover, and now...a cable with three wires is being laid down from Cromer to Helgoland and Denmark, thus giving a more direct communication with Denmark and the north", "Submarine Telegraphs", *Morning Post*, 11 July 1859.

5
The Triumph of the Oligopoly

The early 1860s flowed on seamlessly from the previous five years. Despite the poor performance of the duopoly and its alienated users touting the entry of any rivals bent on destroying it, the telegraph market carried on as before. The Electric and the British & Irish stood compact against the danger of their competitor, the United, which proposed a flat charge on the lines joining the main economic and political centres of the country. However, the United soon changed tack and went in with the duopoly, which was consequently enlarged into an oligopoly with price cartels being openly adopted. The consolidation of this situation between 1855 and 1865 was complementary to the prodigious development of British submarine telegraphs, which, based on great investments, scraped together capital, management and know-how from landline telegraphs, furthering the strategy of scarce investments and high dividends common to the companies of landline oligopoly. Their triumph in 1865 brought to the surface problems that had been upsetting users since the beginning of the decade, who now moved from being upset to openly and fiercely critical and ended by causing a drastic change in regime.

5.1 Challenging the duopoly

Midway through the 1860s, the discontent of the users was catalysed and exploited by the entry of the United Kingdom Electric Telegraph Company (the United), which took to the field with an offer of exactly what the users wanted most, a universal service offering a cheap tariff with a flat charge irrespective of distance.[1] Following the example of the Penny Post's costing policy, the United had introduced a popular one-shilling tariff and had brought in clients who up to then had turned to it only on rare occasions.[2] In the same way, extending the service to new users would bring about a considerable increase in traffic and therefore income.[3]

---

[1] "We're not about to complain of the telegraphic companies which have thus far had the whole business of the country in their hands. They have had many difficulties to overcome and have conferred immense advantages upon the public. Still, the time has now arrived when it must be made known that telegrams in this country cost much more than they ought not chiefly on account of the large profits made in telegraphic apparatuses have put into our hands the means of doing what those companies, with their old and costly contrivances, cannot afford to do. ... What we require is to commence a new and improved telegraphic system *ab initio*, so that we may give the public the benefit of all the great advances which have been made in the construction of telegraphic instruments. Ever since the telegraph has open to the public, the number of messages sent by it has been, as we might have predicted, continually on the increase. If we can succeed in lowering considerably the present high price of the telegram, who can doubt that the number would be correspondingly multiplied? The postal system, which is of a character essentially similar to that of the telegraphic system, shows us what we may fairly anticipate. Since the cost of postage was made low, and uniform throughout the kingdom, the number of letters sent by post annually has been increased nearly seven-fold... Now, what we desire is to imitate this progress (proportionately) in our telegraphic system; and to do this it is only necessary, we are persuaded, first to make the cost of a message low, say a shilling for fifteen words, and next, to make this cost uniform for messages sent to any part of the kingdom. ... Now the question arises – can a telegraphic system be so established throughout the United Kingdom as to render a uniform shilling message remunerative? We answer – without doubts. But such a system must embrace many novelties. ... The person to whom we are indebted for the latest and most important improvements in electric telegraphs...is Mr Thomas Allan". "The Progress of the Electric Telegraph", *Caledonian Mercury*, 23 July 1860.

[2] "By the adoption of a *universal single rate of charge*, on the admirable principle of the penny post, not only will existing telegraphic Business be greatly augmented, but a new class of customers will be called into existent. Tradesmen requiring the immediate supply of goods will find it advantageous to spend one shilling, where the margin of profit on orders renders the present rates prohibitive. The Simplicity and uniformity of charges, will greatly induce persons to telegraph, who are at present deterred by the uncertainty of the cost, as well as by its amount". BT, *United Kingdom Electric Telegraph Company*, TGH 2/1/1, *Prospectus 1860*.

[3] "It is worthy of notice that the reason alleged for the maintenance of the existing excessive charges is, that the wires are even now fully occupied with messages, and that a reduction could not therefore be advantageously made. The object of this Company will be not to retain high charges and limit communication, but to lower Telegraph rates, and make provision for the largely increased traffic inevitably resulting from such a measure. Its object is to popularize and bring the Telegraph within the reach of the great mass of the public. This it feels satisfies of accomplishing", Ibid.

In reality, the United had already tried to enter the telegraph market back in 1851, when it obtained parliamentary authorization for the commercial exploitation of the telegraph invented by Thomas Allan, the company's main promoter. He had inherited from his father Thomas Allan & Co., a printers and publishers also responsible for the *Caledonian Mercury* and the *Encyclopaedia Britannica*.[4] In 1853 Allan persuaded a group of respectable businessmen to come on to the board of the United[5] and in his function as superintending engineer made public its financial prospectus. The main features he published were the introduction of a new signalling system, the adoption of an innovative telegraph apparatus and the use of a more economical insulating system.[6] A price reduction and the possible extension of a flat charge were ventilated as possible consequences of the radical improvements brought about by the three innovations too.[7] Despite a vigorous promotional campaign by means of his own newspaper too,[8] Allan did not convince any shareholders to adhere to his initiative and had to abandon it.

---

[4] All the biographical matter on Thomas Allan comes from Steven Roberts, "Thomas Allan", History of the Atlantic Cable & Undersea Communications, 2010 (updated 5 January 2015), available at http://atlantic-cable.com/CablePioneers/Allan/, accessed on 3 January 2015.

[5] "In 1853 the directors were: Edward Barnard, director of the Bank of Australasia, David Barry, English director of the Paris and Orleans Railway and its Extensions, Charles Cowan, M. P. for Edinburgh, Samuel Daniel, Robert Gillman, director of the Swedish Railway, Alexander Hastie, M. P. for Glasgow, Thos. Wingate Henderson, director of the Edinburgh and Glasgow Railway, Charles Joyce, Standish Motte, James Pilkington, M. P. for Blackburn, James Pillans and Fred G. B. Ponsonby, chairman of the Dublin and Wiclow Railway Company. BT, *United Kingdom Electric Telegraph Company*, TGH 2/1/1, *Prospectus 1853*.

[6] "*First.* – A better system of signals, by which a greater number of dispatches can be transmitted in the same time than by the existing arrangements. *Secondly.* – A variety of improvements in the Telegraphic apparatus, having for their principal object to make the instruments act at far greater distances than has been hitherto attained, and to work a greater number of instruments simultaneously in the same circuit. Thus enabling a despatch to be telegraphed without break or repetition a distance of 1000 miles and upwards. *Thirdly.* – A more economical method of insulating and protecting the conducting wires, especially for submarine Telegraphs", Ibid.

[7] "It is proposed to lay down the main lines (subterranean) of a complete system of Telegraphic Communications throughout Great Britain and Ireland, with a sufficient number of auxiliary branches to establish an immediate correspondence between all the principal towns, and to provide for its subsequent progressive development. There can be no question that the adoption of a small charge, say 1s. per message, or 1d. per word, would be amply remunerative, and as the business of the Company extended, produce a large revenue", Ibid.

[8] "United Kingdom Electric Telegraph Company", *Caledonian Mercury*, 23 May 1853.

Unexpectedly, the project came to life again in 1869 when a group of possible financers grouped around Allan, who was coming out of seven years of business failures in the field of telegraphs and electricity. The financial prospectus, which began circulating in July 1860, set its main aim as introducing the flat charge of a shilling[9] and unlike the original project, put Allan's telegraph[10] into the background. In addition, the 1860 board contained only three directors from 1853: Edward Barnard, manager of the Bank of Australasia, Charles Joyce, on the board of the Bank of London, and James Pilkington, a member of Parliament for Blackburn. Other figures from political and financial circles joined them: Sir Henry Leeke, admiral and member of Parliament for Dover, who was elected chairman, Lord Alfred Churchill, member of Parliament for Woodstock, Angus Croll, ex-manager of the Great Central Gas Consumer Company of London, James Nugent Daniell, chairman of the Blackwall Railway Company, Frederick Doulton, of the famous pottery family, George Braginton, a banker, and Colonel William, Elsey, another Bank of London director.[11] Probably this group of dealmakers had sensed the potentiality of a telegraph company challenging the duopoly by offering lower tariffs.

To accelerate the process these figures had associated with Allan, who for years had been searching for financial partners, but more interestingly had already obtained parliamentary authorization. This was why Allan was initially presented as an electrician and engineer,[12] a position he lost almost immediately once the inefficiency of his telegraph was ascertained. At this point the board offered Allan the possibility of staying in the company with a far

---

[9] "The object of this Company is to establish a system of Electrical Communication, based on the principle of the Penny Postage, to convey messages throughout the United Kingdom at a low and uniform rate, a shilling for a short message, or some equivalent charge, irrespective of distance". "The United Kingdom Electric Telegraph", *London Evening Standard*, 26 July 1860.

[10] "The system of Telegraphy hitherto in use in this country are inefficient to accomplish a large amount of traffic in a given time, besides being needlessly expensive, owing to the requirement of so much skilled labour in their operation; whereas, under Allan's system, by the introduction of labour-saving apparatus, Telegraphing will be reduced to as mere mechanical operation requiring no skilled labour, obviating error, and obtaining the maximum of speed with the minimum of cost, thus producing traffic and profits in a progressive ratio", Ibid.

[11] Other board members were Robert Bryce Hay and Edward W.H. Schenley. Ibid.

[12] BT, *United Kingdom Electric Telegraph Company*, TGH 2/1/1, Prospectus 1860.

lower salary plus a one-off payment.[13] He rejected the offer and went in for a public battle from the pages of his daily paper implicitly in search of a handsome indemnity.[14]

Like the British in the early 1850s, the United adopted right from the beginning an aggressively competitive strategy against the

---

[13] "It was considered so great matter to accomplish this as to entitle Mr. Allan to the thanks of the Directors; it would seem, however, that at the very time they were publicly complimenting him on his exertions, and entering into contracts to carry out his plans, they were privately meeting and demanding a reduction on the money terms of the original agreements with him, and indeed manoeuvring to get him out altogether. The Board met on the 2nd October and resolved that the agreement of the 24th July on which the Company was based, should be cancelled '*so far as the cash arrangements go*' and following, a great reduction, should be proposed to Mr. Allan in lieu of his previous settlement: – 'CLAUSE I. That the salary should commence at £ 600 a year. CLAUSE II. That the royalty should remain as before, except that is continuance should be made contingent on the Board finding it consistent, with the interests of the Company, to continue to use Mr. Allan's Patents. CLAUSE III. That the bonus to be given to Mr. Allan shall be £ 12.500, half in cash, half in shares. The board have the power to make Mr. Allan any further remuneration should they desire or deem it necessary, when the capital of the Company shall exceed the sum of £ 150.000'. The agreement thus worded, and reluctantly consented to by Mr. Allan, was the next day drawn out and with it the following: – CLAUSE IX. The Directors of the Company and the said Thomas Allan may, at any time or times, make such alterations and modifications in this agreement *as may be mutually agreed upon*. Provided always, that nothing herein contained shall be constructed to make it binding or obligatory upon the said Company, or the Directors thereof, to use all or either of the said Patents, in case they shall, in their absolute discretion, determine at any times not do so; and upon notice of such determination being given to the said Thomas Allan, or left for him at his known place of abode in England. This agreement, and every clause, matter, and thing herein contained, shall absolutely cease and determine; and the said Company, or Directors thereof, shall not in any way be answerable or liable to the said Thomas Allan for anything done under or by virtue hereof in reference to the said Patents or otherwise be witness whereof", *Caledonian Mercury*, 11 November 1861.

[14] "The agreement, however, in this equivocal, if not fraudulent form, was submitted to Mr. Allan, who naturally enough was profoundly astonished at it, and indignantly resented it, remarking among other things: – 'Looking at the present aspect of affairs, I don't see how we can proceed upon the amicable footing we should do, and great alterations must take place to make them better; but to show how much I desire to see them prosper, I will either retire or remain. My terms of retirement shall not be mercenary ones. I wish, however, to state, that, whatever position things may come to, the agreement upon which the constitution of the Company is based is the one upon which I desire to adhere. I will not make the pecuniary engagements a *sine qua non*. I am ready, in some way to modify this, and if it is arranged, I should retire, have a sum named as a liquidation and in full of all demands'", Ibid.; "My statement of facts addressed to the original subscribers to the United Kingdom Electric Telegraph Company, drawn forth by peculiar actions of the directory, and which you have so largely quoted and commented on, cannot be confuted by any one of the Board, and must be my reply to all his Lordship's insinuations. I will therefore, with one remark, not intrude myself further, but with confidence leave his Lordship and his confederates to the tender mercies of public opinion", "The United Kingdom Electric Telegraph Company", *Caledonian Mercury*, 19 November 1861.

duopoly, based mainly on anti-monopolistic propaganda, particularly appreciated by businessmen and the press. This was why the press gave their unconditional support to its entry, just as had happened with the British,[15] and often described it as the only chance to save British telegraphs from the stranglehold of the two monopolies.[16] Yet again the Electric is identified as the main monopoly in both England and Scotland, side by side with the British & Irish, which held Irish telegraphs in its grasp.[17] In introducing a flat charge between the main cities in the country, the United would force the

---

[15] "The works of the United Kingdom Telegraph Company between London, Brighton, Manchester, and Liverpool have recently been pressed forward with considerable vigour, and the lines are now progressing on several points simultaneously. This is the company, it will be remembered, which proposes to carry out all over Great Britain a system of uniform charge for telegraphic messages. It is intended to fix the amount at 1s. so as, by the adoption of a low rate, to encourage the public to use the telegraph, and at the same time to avoid the uncertainty which is at present felt on the mileage system as to what the charge for a message may really be", "Cheap Telegrams", *Morning Post*, 26 February 1861.

[16] "The two existing companies – the one secure in its monopoly of the Irish lines; the other guarded by its monopoly of the English and Scotch ones – could not be expected to lower their rates of charge, and at length the idea (proposed by Mr Thomas Allan of Edinburgh) was suggested of erecting pole telegraphs along the great canal routes and the lines of turnpike road. Here is the true solution of the problem. The lines can be erected as safely and as cheaply; the durability is equally secured; but Parliamentary powers must be granted for the purpose. The public call loudly and incessantly for cheap telegraphs. Cheap telegraphs must necessarily be pole telegraphs. The canals only secure communication here and there. The public demand the telegraph everywhere, and the public roads of the kingdom alone afford the necessary routes, and must step in to save the country from the everlasting incubus of dear telegraphy. We are no supporters of individual companies, but if the United Kingdom Company introduce and carry out its system along the canals and roads of an uniform shilling rate, irrespective of distance, it will deserve well of the country", "Why Telegrams Are Dear [from Observer]", *Caledonian Mercury*, 26 March 1861.

[17] "Our Readers, in common with those of other daily papers of this city, are not ignorant of the generally meagre and blundering character of the telegrams transmitted by the Electric Telegraph Company to the Press; unless however, they had the experience of those whose duty it is to make the best efforts to correct them in MS., they could form no conception of the style in which many of the messages are accustomed to be sent to the various offices. Though the company here have largely increased their terms of supply to the Press, the news forwarded is oftentimes the most meagre and imperfect, frequently about matters of no earthly interest to the Scottish public, information of special interest being neglected, and at times considerably after date, important items of intelligence which have appeared in the London evening papers not being transmitted till the following morning. The company has in the meantime a monopoly, and I therefore at liberty to do just as it likes. It would seem from the following, which we clip from the *Northern Whig*, that the Magnetic Telegraph Company, also a monopoly so far as Ireland is concerned, is conducting itself in a similar manner there", "The Press and the Telegraphs", *Caledonian Mercury*, 25 August 1860.

two rival companies to change their tariff policy, to the benefit of all.[18] Nevertheless, the rivals reacted to the United's press campaign with an aggressive commercial and legal battle. First of all, when the United's prospectus began to circulate again in July 1859, the Electric began to question the validity of its parliamentary authorization.[19] According to the Electric's lawyers, the 1851 authorization referred to a stock company with unlimited liability, without capitalization. Later, however, two regulations for the establishments of stock companies in 1855 and 1856 revolutionized the minimum requirements for a company of this kind. According to the Electric, the United as advertised in 1860 was formally different from the 1853 version, and as such could not enjoy the same privileges.[20] This was why the United still claimed its rights according to the 1851 authorization,

---

[18] "It is now an established fact that a decreased tariff means an increased traffic. Lower the duty on any one article of consumption and the increased demand for it will more than compensate the revenue for the reduction of duty. ... If the United Kingdom Electric Telegraph Company only succeed in carrying out their proposed system of one shilling messages to all parts of the United Kingdom, they will not only confer an invaluable boon on the public, but will ensure to themselves (what, no doubt, will be more gratifying) a very large and handsome dividend for their outlay in doing so. Their success will inaugurate a new era in electric telegraphs nearly equal to the benefits the public have derived from the penny postage", "Cheap Telegraphs", *Morning Chronicle*, 2 August 1860.

[19] "The Chairman – ... with respect to the Company to which the honourable proprietor alludes, it has, of course, been brought before our notice through the medium of the press. And I had no wish to have said one word about it, notwithstanding the many misrepresentations which are contained in the prospectus, and also in a pamphlet which accompanies that prospectus. We have been advised that the Company which is proposed under the title of 'The United Kingdom Electric Telegraph Company' has no parliamentary powers whatever. In its prospectus it says it is empowered by special Act of Parliament, 14 & 15 Vic. Cap 137. I have, however, been advised that they have no parliamentary powers, and that I should be very unwilling not to do so", BT, *Electric Telegraph Company*, TGA 2/1, *Half Yearly General Meeting*, August 1860.

[20] "To the editor of the daily news. The Electric and International Telegraph Company, Telegraph Street, London, E. C., Aug. 3, 1860. Sir – The statement of the chairman of this company made at the late general meeting, in answer to a question from a proprietor, to the effect that the United Kingdom Electric Telegraph Company (Limited) is not empowered by the special act of 1851, having been challenged, I am instructed to send you copies of a case and opinion of counsel upon which that statement as founded. ... J. S. Fourdrinier, Secretary. In December 1850 a company was provisionally registered under the name of 'The United Kingdom Electric Telegraph Company'. An act of Parliament sent herewith was passed in the session of 1851, granting powers to this company when it should be completely registered. The company issued in 1853 a prospectus, sent herewith, but it has never been completely registered, and in consequence of the lapse of time and the changes which have since taken place in the law its complete registration is now impossible. In the present month, July 1860, a prospectus

but also obtained the extension of its powers,[21] guaranteed by a second parliamentary procedure.[22]

Secondly, the Electric and the British & Irish together possessed the exclusive rights for the installation of telegraph lines along all the railroads and highways in the country. As we have explained, no company could build overhead lines along ordinary streets and roads without special permission, which was why the United was in 1861 forced to sign wayleave contracts with the companies running the canals,[23] a wasted effort, because in the meanwhile the United had obtained a new parliamentary authorization for installing lines

---

has been issued, announcing the formulation of a limited company by the name of 'The United Kingdom Electric Telegraph Company (limited)'. This prospectus is sent herewith, and counsel is requested to observe the statement with which it commences, that the company now established is empowered by the special act of parliament of 1851 above referred to. ... Whether it is not clear that the company now formed has no title to the powers conferred by the special act of 1851, and whether the statement of its prospectus claiming parliamentary powers is not clearly contrary to the fact? OPINION. The special act has no application whatever to the company later formed. This company has not in my opinion any manner of identity with the projected company which provisionally registered. It necessarily follows that the assumption by the new company of the rights or powers conferred by the special act of 1851 is wholly unwarranted", WM. Cracroft Fooks, *London Daily News*, 4 August 1860.

[21] "The following letter from the secretary of the United Kingdom Electric Telegraph Company (Limited) Gives a complete denial to a statement made by the chairman of the Electric and International Telegraph Company at the meeting yesterday: 'TO THE EDITOR OF THE DAILY NEWS Sir, – The extraordinary assertion by the chairman of the Electric and International Telegraph Company, as reported in several of the morning journals of this day, namely, 'that the United Kingdom Electric Telegraph Company has no parliamentary powers' is directly untrue; this company having the benefit of special and ample powers conferred by the 14th and 15th Vic, cap. 137, which act is now lying at this office, for the inspection of the public. ... John Lilwal, Secretary, The United Kingdom Electric Telegraph Company (Limited), 101, Gresham House, London, August 2, 1860'".

[22] *London Standard*, 2 February 1861.

[23] "The Shareholders will remember that doubts were thrown upon the Act of Parliament (14 & 15 Vic, c 137) conferring underground Powers on high roads upon the first introduction of the Company. The directors immediately thereupon received information from quarters upon which reliance could be placed, by which they were institute full enquiries and collect varied evidence as to the best and most remunerative systems of telegraphy; and the result of these enquiries was decidedly to establish the advantage of lines on poles as against lines underground on the high roads. The Directors therefore entered into negociations [*sic*.] with the various Canal Companies, and with the Trustees of the Turnpike Roads, for powers enabling them to construct lines of pole telegraphy; and when these arrangements were sufficiently matured, they issued, with the letter of allotment, a circular explaining to the Shareholders the altered plans and position of the Company", BT, *United Kingdom Electric Telegraph Company*, TGH 2/1/1, *Directors' Report*, July 1861.

along the roads.[24] It was still forbidden however to cross railroads, so that the railway companies, backed by the Electric and the British & Irish, had the United's lines pulled down whenever and wherever they crossed railtracks.[25]

In the third place, the Electric and the British & Irish always declared they were against the United's tariff plan, accusing it of being financially unsound.[26] The experience of the two companies showed in fact that applying a flat one-shilling charge would

---

[24] "Although the Company's Bill, was hotly contested at every stage, and although efforts were made to thrust in objectionable clauses, these efforts were frustrated in every case; the Electric and International Telegraph Company especially exhausted every argument without success, and the United Kingdom Company finally gained their Act; and are now in a position to erect and maintain Telegraphs all over the Kingdom, without incurring rentals to Canal and other Companies, and with such advantages as cannot fail to secure success", BT, *United Kingdom Electric Telegraph Company*, TGH 2/1/1, *Directors' Report*, July 1862.

[25] "Francis Lamb and Thomas Whitfield, two inspectors in the employ of the London and North-Western Railway Company have been convicted before the magistrates at Woodstock, of unlawfully and maliciously destroying certain telegraphic wires belonging to the United Kingdom Electric Telegraph Company, at the level crossing on the Banbury turnpike road. The case was clearly made out and they mere committed for trial at the next quarter sessions, to be held in January next. Mr. Hill, barrister, admitted that the London and North Western Railway Company instigated the defendants to cut the wires, and sent an engine with men and ladders to enable them to do so, as they maintained they has a right to prevent anything being carried across their line. Mr. Atkinson, on the other hand, contended that the railway company had no right to interfere with any wires crossing a turnpike road where the trustees of the road had given permission for their erection, and that as the wires were suspended 35 ft. above the road, it could not cause obstruction to any traffic either on the rail or on the road. It appears that the railway company is mixed up with the International Electric Telegraph Company, who are [*sic*] opposing the United Kingdom Company, which was the first to introduce a uniform rate of 1s. for telegrams", "Destroying telegraphic wires", *North London News*, 14 December 1861.

[26] "This company has come before the public, and holds out to them the boon of a uniform tariff throughout the country at a very low rate; and it asserts that it can do this by means which are not at the disposal of any other Company. It first speaks of giving direct circuits to all principal commercial towns in the country. There is nothing novel in that, for our Company has been doing that many years past. Their second mode is in working for the press, in sending the same intelligence to several towns simultaneously – that is a system which we have long adopted – it is technically called working Y. Q. Again, their working long distances by means of relays is a system done by all companies. The fourth is Mr. Allan's Labour Saving Apparatus. Now there are a great many scientific and ingenious gentlemen who have gone to great pains for the purpose of introducing mechanical instead of manual labour. We have tried these inventions, but they have failed; they have failed because they cannot be depended upon, and they are not used by us or any existing companies. 'Labour Saving Apparatus' might more properly be called 'Error Creating Apparatus'. Now gentlemen,

not cover the running costs.[27] It would cause such an increase in traffic and hike running costs, in terms of more personnel, more telegraphs and the installation of more wires. However, in order not to be excluded from the market between the main cities, the Electric and the British & Irish agreed to adopt simultaneously the same tariff as the United's on the circuits in direct competition.[28] So, though doubting the profitability of the flat charge, the

---

we come to the figures – they take 12 of the large commercial towns in this country, and they estimate that those 12 commercial towns will create in round numbers 5,000,000 of messages in the course of a year. We with our 450 stations (including these twelve) have a little more than 1,000,000 a year. ... The whole thing is quite full of absurdities, and it is evident that this prospectus has been penned by some person quite ignorant of practical telegraphy", BT, *Electric Telegraph Company*, TGA 2/1, *Half Yearly General Meeting*, August 1860.

[27] A few rare sceptical journalists trusted the experience of the Electric and doubted the United's strategy: "The Company [United] have adopted Mr. Allan's system of telegraphy, and propose to convey messages, throughout the United Kingdom, at the low and uniform rate of one shilling per message. Great as is our desire to see the use of the telegraph extended, we very much doubt whether a new company, on the limited liability system, possesses greater facilities for producing that benefit than the Telegraph Companies already in the field. Moreover, from what we know of Mr. Allan's instrument, we think it hardly the one to contribute to the desirable end. It is many years since we saw it, but if our recollection serves us right, it is not only inferior in accuracy and speed to the admirable inventions of Professors Morse and Wheatstone, but is even far below the system adopted by the Magnetic Telegraph Company. We have frequently had occasion to complain of the Conservative tendencies of the Electric and International Telegraph Company, but justice compels us to recognise the great services which that Company have rendered to the public. It was the first Company to enter the untrodden path of electricity as a means of communication; and though slow to give the public the benefit of reduced rates, it has steadily gone on improving its system, and perfecting its means of communication. With such a formidable competitor already occupying a high place United Kingdom Telegraph Company are very small. If a low and uniform rate of telegraphy is to be reached within a moderate lapse of time, we cannot but think that a corporation, such as the long-established Company alluded to above, has better means and appliances to arrive at such a desirable consummation, than a new company, which has all its lines to erect. If however, Mr Allan and his colleagues think they can command success, let them go on and prosper", "Electric Telegraphy", *Dundee, Perth and Cupar Advertiser*, 14 August 1860.

[28] The letter that Fourdrinier sent to the United was unequivocal: "I send you inclosed [*sic*.] a Circular which has been spread by the United Kingdom Company. It mentions that their 3 stations in London, and their station in Oxford are open, but we have ascertained that they have also commenced business at Birmingham and Banbury. We have therefore instructed our Clerks to take messages at those stations at the same rate namely one shilling for 20 words between the above stations, and preparing hand-bills", BT, *United Kingdom Electric Telegraph Company*, TGH 2/2/1, *Letter of Mr Fourdrinier to United Kingdom Electric Telegraph Company*, 3 October 1861. Just as explicit are two articles about the tariff reductions of the Electric and the British & Irish: "We have information from

duopoly accepted the challenge, well aware that they could cover any losses on the circuits between the main cities, given the higher tariffs.

Theoretically, the United was helped by parliamentary approval of the first structural law governing the telegraph service. Legal authorization had originally been given to the establishment of, and modifications to, the trade names of the telegraph companies desiring to operate in the United Kingdom. Consequently, up to 1863 the service was regulated by a series of laws which provided for the rights and duties of the individual companies. For the first time a regulation guaranteed any company entering the sector the possibility of building telegraph lines alongside, above and below highways, railroads and canals without needing specific permission. In addition, the 1863 Telegraph Act laid the basis for free competition, putting under strict control transfers, mergers and sales between telegraph companies. Its most important norm was in Article 43: "The company shall not sell, transfer, or lease their undertaking or works, or any part thereof, to any other company or to any body or person, except with the consent of the Board of Trade previously obtained for such sale, transfer, or lease".[29]

The press, however, accused Parliament of having approved a law that was limited to confirming the status quo. First of all, all the foreseen rules, including the one about building lines, had already been applied for many years by the companies which had entered the market before 1863 and had specific authorization. Secondly, Article 43 protected operations transferring property, but stopped short of legislating on cartel agreements, which were exactly what held up the Electric-British & Irish duopoly. So though the Telegraph Act facilitated the entry of new companies, it did not increase free

---

Mr. Wade, the manager of the Electric Telegraph Company, of a reduction having been made in the company's tariff between Birmingham, Dudley, Dudley Port, Great Bridge, Perry Barr, Spon Lane, Walsall, Wednesbury, Bilston, & Wolverhampton, to one shilling for twenty words", "United Kingdom Telegraph Company", *Manchester Times*, 20 April 1861; "The British and Irish Magnetic Telegraph Company give notice, that on and after Monday, the 11th March, 1861, the Tariff for messages to the Company's Stations in Great Britain will be greatly reduced". "British and Irish Magnetic Telegraph Company. Reduction of Rates". *Northern Whig*, 27 March 1861.

[29] *An Act to regulate the Exercise of Powers under Special Acts for the Construction and Maintenance of Telegraphs* [28 July 1863], 26–27 Victoria, c. cxii, Article 43.

competition because it did not question the mechanisms underlying the oligopolistic regime. The absence of any debate over the regulation and the substantial inconsistency of the few amendments proposed proved yet again Parliament's lack of interest in regulating the telecommunications market.[30]

Yet the 1863 Telegraph Act protected the Universal Private Telegraph Company, which had been recently established and could not come under Article 43. Born in 1860,[31] this company aimed at building, renting, running and selling private telegraph lines in London and other cities.[32] It received parliamentary authority in 1861 and was to deal exclusively with a private service, thanks to the introduction of Wheatstone's ABC telegraph, which used alphanumeric characters directly and did not require staff to be specially trained.[33] As the Universal did not carry out any public service, it never entered into competition with the other companies and stayed in a market niche that up to then had been exploited only marginally by the Electric and District.[34]

---

[30] "This bill [Telegraph Act 1863] starts off with the pretence that it is intended to regulate the exercise of powers under special acts for the construction and maintenance of telegraphs. An examination of its provisions, however, leads us to no other conclusion than it is the handicraft of those special companies, and if not drawn up by them or their agents, at all events connived at and approved of and supported by them. How otherwise such a bill could pass its second reading in the House of Commons, and exhibit so few traces of the file in committee, is simply inconceivable. The amendments indeed, inserted in committee are mostly conspicuous by their absence, for, after all, they are merely omissions, and of the smallest account possible. So insignificant, in fact, do we deem them, excepting one of 30.000 instead of 10.000 inhabitants, in the seventh clause, which, however, is not an 'amendment' that this notice is all we mean to award them, and we shall now proceed to criticise the original act, just as though no such amendments had ever been suggested. ... Now we should be glad to know what is the purpose of The Telegraphs Act of 1863. Either it is nothing more than an unsubstantial shadow, or else its avowed object and meaning are a complete misnomer", *London Standard*, 18 March 1863.

[31] "Universal Private Telegraph Company", *London Standard*, 22 December 1860.

[32] "The Universal Private Telegraph Company, whose operations have been carried on so successfully in London, are beginning the erection of poles and wires in Glasgow, thus enabling manufacturers and others, whose works are some distance out of town, to communicate with their places of business or their dwellings in Glasgow. Two lines – one to Dalmarnick and the other to St Rollox – are already completed; and the cable will be suspended, and the private instruments put into working order, in the course of next week. The company have had but little difficulty in getting permission to erect their poles on the roofs of the houses in Glasgow", "Private Telegraphy", 24 October 1861.

[33] *London Standard*, 2 February 1861.

[34] Roberts, *Distant Writing*, pp. 129–133.

5.2   The consolidation of the oligopoly

The dumping strategy used by the United was risky and not well synchronized. The new company may well have been intending to offer a telegram at such a low price as to stimulate a rapid increase in receipts. With the publicity offered to this success, the clients of the Electric and British & Irish would be induced to change over to the new company.

Once the duopoly was in real trouble right in the most profitable circuits, the United would be able to negotiate on favourable terms with the Electric and British & Irish and enter an oligopoly which for years had guaranteed high profits. If in theory the tactic might have worked with a good dose of fortune and technological means the United did not possess, it worsened the situation by announcing the new tariff well before its line began operating. The flat one-shilling rate was announced and widely advertised already in July 1860, before the company had begun to build the lines. The later delays caused by their rivals' legal opposition, which forced the United to open negotiations with the canal companies and apply for a new Parliamentary authorization, meant that the first telegrams could not be sent until November 1861. In the meantime, the Electric and the British & Irish had reorganized themselves and announced their tariff adjustments before the United was operative.

The counter tactic employed by the Electric-British & Irish created serious difficulties for the United, which found itself without any other innovative arm to rely on. In 1862 it tried to introduce some financial and technological novelties. The issue of "frank-messages bonds" turned out to be well chosen. The campaign was directed at all those who needed the telegraph or would have liked to use it mainly for reasons of work. The interest on this stock was paid up not in cash but in bonds, which guaranteed no charge for telegrams sent via the United lines.[35] Some of the company's managers, like Lord Alfred Churchill, took part personally in the advertising campaigns, which were reasonably successful with the middle

---

[35] "With the object of ensuring permanent support for any extensions that might be made, several important telegraphing towns in South Wales were visited early this year, and the merchants came forward and subscribed the amount required to bring down the Company's wires to their towns. So strong was the feeling in those

classes, but as was foreseeable could only guarantee a fairly limited financing.[36]

At the same time, the United tried to introduce Hughes's printing telegraph,[37] the first to print in the letters of the alphabet. This mechanism had the enormous advantage of making it possible to

---

places, that not a merchant telegraphing to the extent of a few pounds a year, but subscribed for the Company's capital, receiving an interest in Frank Messages stamps. Wherever, therefore, extensions are made under these circumstances, the connection and business of the towns are assured to the United Kingdom Company. The amount of interest paid in Frank Stamps scarcely averages a twentieth part of the sum annually spent by such parties in sending telegrams; so that a large revenue is besides obtained by the Company by the connection these stamps secure. In consequence of this success, it has been determined, with the view of ensuring custom to the Company, to offer, in the first instance, the remaining shares to merchants and others who use the telegraph in London and the different provincial towns, or to issue bonds bearing a guaranteed interest payable in the Company's Frank Message Stamps to the same parties. The directors need scarcely observe that this measure is the best calculated to ensure future prosperity, and should of itself alone make the Company a most remunerative investment", BT, *United Kingdom Electric Telegraph Company*, TGH 2/1/1, *Directors' Report*, July 1862.

[36] "The basis on which they were now proceeding was that they should ask the different towns they came to subscribe the necessary amount of money that might be required to bring their lines down there either by taking the ordinary shares of the Company or bonds giving a guarantee interest of 7 ½ per cent, to be paid in the frank message stamps of the Company transferable", "United Kingdom Electric Telegraph Company. Uniform Charge for telegrams", *Caledonian Mercury*, 5 July 1862; "On Monday afternoon, an influential meeting (convened by the Mayor) of mercantile and commercial gentlemen was held in the large room of the Exchange Buildings for the purpose of receiving a deputation, consisting of Lord Alfred Churchill, M. P., one of the directors, and William Andrew, secretary and manager, from the United Kingdom Electric Telegraph Company (Limited). ... They proposed to raise the capital by two methods – first, by the issue of ordinary shares, and next by the issue of debenture bonds, bearing interest at the rate of 7 ½ per cent. To be paid in stamped franks for messages. He had no doubt that gentlemen who telegraphed by the existing telegraph companies were aware that they issued stamped franks and the United Kingdom Telegraph Company only proposed to do the same thing. The holder of these stamped franks would only have to write out his message, on any kind of paper, place the stamp upon it, send it to the office, and it would be at once transmitted, without any money passing between him an the clerk. The interest of 7½ per cent on the debenture bonds would be simply paid in these stamps. Any person so investing his money would receive this interest on his deposit. Such a mode of investment would benefit both the investor and the company, because it would be the means of bringing traffic to the company and of giving a return to the investor in covering his messages. Suppose a gentleman was to invest £ 200, he would receive £15 in telegram stamps as the interest for one year, and would be enabled thereby to frank messages on every one of the three hundred days in the year", "United Kingdom Electric Telegraph Company", *Bradford Observer*, 25 September 1862.

[37] "The Directors, alter experiments carefully conducted, and alter observing the practical Business performances for a whole month of the Hughes's Printing Telegraph

print messages immediately without using conventional codes like Morse. Needless to say, the speed and accuracy of the messages sent and received could be drastically increased by the introduction of the Hughes system; however, it had a significant flaw in that it was extremely delicate, since the principle on which its performance was based was synchronism. The different mechanisms connected with one another had to be perfectly synchronized so that the letters sent would be the same as the ones received.[38] For this reason staff needed a three-month training period, which could be given however only to those who were expert in transmitting telegraphs and had a minimum knowledge of scientific matters. But the United was a new company without expert telegraph operators and could not take them on because it had to contain running costs. Consequently, though it possessed a highly productive telegraph, the United could not exploit all its potentialities or extend it to all its stations.[39]

Though the United directors declared they were optimistic that their innovations would soon generate high profits,[40] the company

---

Instrument, deemed it judicious to secure the patent of this – the only practical rapid printing instrument known. A staff has been educated for the purpose of working this apparatus, and it is now, and has been for a long period, doing the commercial work of the Company on certain circuits with great efficiency and despatch. The Directors anticipate considerable advantages from the application of this instrument, as they are enabled gradually to extend its use; and as the messages are delivered by it in clear ordinary newspaper print to the merchants' offices, the Directors are led to believe that this system of communication will be greatly preferred to that ordinarily in use, not only on account of the message being delivered in a printed form, but also from the fact that errors are rendered almost impossible by the use of this valuable apparatus", BT, *United Kingdom Electric Telegraph Company*, TGH 2/1/1, *Directors' Report*, July 1863.

[38] M. Ch. Bontemps, *Les systèmes télégraphiques aériens – électriques, pneumatiques*, Paris, Dunod Éditeur, 1876, pp. 115–118; Edward B. Bright, *The Electric Telegraph*, London, James Walton, 1867, pp. 159–160.

[39] "The Directors are pleased to state that the staff of the Company having become more settled, they have begun to bring the Hughes Printing Telegraph into daily practical operation between Liverpool and Manchester, where its performances are most satisfactory", BT, *United Kingdom Electric Telegraph Company*, TGH 2/1/1, *Directors' Report*, July 1865.

[40] "The Directors are more than ever confirmed in their opinion that a cheap uniform system will prove ultimately the most remunerative. The revenue of this Company has doubled since the lines were first opened; and this, although nearly the whole of the time of the officers has been taken up in overcoming the opposition got up in the law courts and in Parliament; and although the lines have been undergoing continual alterations and removes; and although the wires of the Company have been malevolently interfered with, with the object of Shopping the transmission of the messages", BT, *United Kingdom Electric Telegraph Company*, TGH 2/1/1, *Directors' Report*, July 1862;

Table 5.1 Accounting data for the Electric 1861–1865 (in £, without the decimals)

	Expenses	Receipts	Balance	Dividend	Fund
June 1861	68,786	102,519	33,733	28,975	6,141
December 1861	73,276	111,053	37,777	28,975	8,801
June 1862	71,132	102,367	31,235	29,815	6,362
December 1862	80,947	117,074	36,127	31,075	5,051
June 1863	78,149	117,210	39,061	33,057	6,003
December 1863	86,582	137,150	50,568	34,120	16,447
June 1864	87,643	136,473	48,830	38,995	9,835
December 1864	87,514	141,590	54,076	39,245	14,831
June 1865	91,041	145,736	54,695	45,556	9,139
December 1865	95,910	168,290	72,380	51,555	

Source: BT, *Electric Telegraph Company*, TGA 2/1, *Balance Sheets* from June 1861 to December 1865.

was always unprofitable, with marginal revenues, despite the notable increase in telegrams.[41] Differently, the Electric and the Magnetic, which followed the same business policy as in the previous five years, continued to chalk up the same rhythm of growth. In the particular case of the Electric, the revenue dipped between 1861 and 1862, just when the new tariff was introduced, and then started to rise again vigorously in the following years and bring up the profits.

In 1864 the United was submerged under its debts, its financial position was unsustainable and its directors realized that in the position they were in their service could never have generated the profits they had originally estimated.[42] At that point they were forced to

---

"Under these circumstances, therefore – and with very enlarged basis on which the Company stands, with one of the richest, most influential, and business-like proprietary bodies in the country – the Directors confidently anticipate at the next meeting of the Shareholders to be in a position to place before them a favourable account of the Company's operations. No exertion will be wanting on their part to secure this desirable end", BT, *United Kingdom Electric Telegraph Company*, TGH 2/1/1, *Directors' Report*, July 1863.

[41] BT, *United Kingdom Electric Telegraph Company*, TGH 2/1/1, *Balance Sheets*, from June 1861 to June 1865.

[42] "Seeing that the Company was working under the greatest possible advantages under these circumstances, and that upwards of four years had elapsed since the formation of the Company without the payment of any dividend to the Proprietary, the Directors conceived that they would not be justified in continuing the shilling system, and arrangements were, therefore, agreed to for its alteration. The Directors waited until

turn to the Electric-British & Irish to try to find an agreement to avoid going to the wall. Half way through 1864, the top levels of the three companies met several times to examine the situation carefully and identify possible solutions.[43] On 23 September 1864 the United sent an official letter to the other two companies giving the five cardinal points for constructing a cartel agreement. Firstly, 12 months were established as the time limit in which the three companies would have to work out together a new common tariff to replace the flat charge on the circuits where it had been offered between 1861 and 1864.[44] Secondly, the United would undertake to avoid any expansion of its network in the following three years with the exception of some previously scheduled lines.[45] Thirdly, the three companies would support one another in transmitting news along the circuits which up to then had been in direct competition.[46] Fourthly, the

---

the last moment before reluctantly adopting this step, but having sought publicity in every way, having persistently canvassed in every department of business, and having endeavoured by the personal solicitations of numerous active agents, to attract trade, they at last saw themselves compelled to agree to a measure that was greatly antagonistic to their personal wishes, but absolutely essential for the well-being of the Company and requisite, as they believe, for the permanent interests of the telegraphing community....The Directors have acted, to the best of their belief, for the permanent advantage of their Shareholders, Frank-stamp Bondholders, and the public; for with an unremunerative tariff the Company could not have been permanently established, whilst the action taken ensures a large present reduction upon former rates to the telegraphing public, and a fair prospect to the Proprietary of a moderate return for their investments", BT, *United Kingdom Electric Telegraph Company*, TGH 2/1/1, *Directors' Report*, July 1865.

[43] "Gentlemen, Conversations having taken place between me and the Secretary of the Electric and International Telegraph Company and subsequently between Mr. Weaver and the Secretary of the British and Irish Magnetic Telegraph Company, and subsequently again between Mr. Weaver, Mr. Bright and myself, and certain bases of a proposed arrangement between the three Companies having been formulated, and now, with the authority of my Board have the honour to propose", BT, *United Kingdom Electric Telegraph Company*, TGH 2/2/1, Letter from the *United* to *Electric* and to *British and Irish*, 23 September 1864. The letter in the archive is neither signed nor dated, probably because it is a copy, unsent. However we can identify it as the letter sent from United to the other two companies on 23 September 1864.

[44] "That a period shall be fixed within which the rate now charged to the public for the transmission of telegraphic messages shall be increased...That this period shall be fixed at 12 months dating from the completion of this arrangement", Ibid.

[45] "That the United Kingdom Company shall understand not to extend over term of three years their operations to any additional terms, or places outsider their present system, beyond or except, certain lines to be scheduled", Ibid.

[46] "That the British and Irish Magnetic Telegraph Company Limited and the Electric and International Telegraph Company shall give to the United Kingdom Electric Telegraph Company an equal share of the news Business in those town which the

United would have to bear the direct costs of recognizing publicly that it was responsible for the failed tariff policy that the companies were now forced to abandon.[47] Lastly, the three companies would decide together on a new tariff which would be introduced simultaneously on their circuits.[48] The official answer to the United was therefore only a formality, sanctioned by two identical letters sent by the Electric and the British & Irish with a text that had evidently been organized in the previous weeks.[49]

As foreseen in June 1865, the boards of the three companies carried out a dense correspondence defining the details of the price cartel,[50]

United Kingdom Company offered on condition that the United Kingdom Company bear their proportion of the expense of providing the news and share the work in their proportion to the press, news rooms, and that they do not compete in the rates", Ibid.

[47] "It must be distinctly understood that as the United Kingdom Company will have failed to make a uniform shilling rate remunerative if the rate be raised, the United Kingdom Company shall publicly in each town where they have a station declare by printed Bills or otherwise that the shilling rate has proved a failure", Ibid.

[48] "The three companies shall in this case simultaneously raise their rate to the same amounts for the same distances so that the rate charged may be identical, and it be suggested that the circular in printed Bills to be offered by the United Kingdom Telegraph Company", Ibid.

[49] "Gentlemen, in answer to the letter of the Secretary of the United Kingdom Telegraph Company dated September 23rd, containing a proposal, on your behalf, that the rates now charged to the Public for the transmission of telegraphic messages, between certain places, should at some future period be increased, I am instructed to say, that my directors have no desire to maintain a tariff which has been proved to be detrimental to the property entrusted to their care. They are willing therefore at the period named by you, to meet the views of your Board in adjusting a tariff which would give satisfaction to the Public, and at the same time, afford a fair and moderate remuneration for the service performed. I am Gentleman your obedient servant Weaver, Secretary", BT, *United Kingdom Electric Telegraph Company*, TGH 2/2/1, Letter from Electric to United, 11 October 1864. "Gentlemen, in answer to the letter of the Secretary of the United Kingdom Telegraph Company dated September 23rd, containing a proposal, on your behalf, that the rates now charged to the Public for the transmission of telegraphic messages, between certain places, should at some future period be increased, I am instructed to say, that my directors have no desire to maintain a tariff which has been proved to be detrimental to the property entrusted to their care. They are willing therefore at the period named by you, to meet the views of your Board in adjusting a tariff which would give satisfaction to the Public, and at the same time, afford a fair and moderate remuneration for the service performed. I am Gentleman your obedient servant E. Bright, Secretary", BT, *United Kingdom Electric Telegraph Company*, TGH 2/2/1, Letter from British and Irish to United, 11 October 1864.

[50] "A Communications from the Secretary of the United Kingdom Electric Telegraph Company containing a proposal to increase the present rate for Inland messages as follows vic. 100 miles 1/- 200 miles 1/6 over 200 miles 2/- was read. Resolved that the proposal of the United Kingdom Company be accepted, it being distinctly understood that the United Kingdom Company have failed to make an uniform 1/- rate

made official 10 June. From then on the three companies would apply the same tariff based on the distance covered by the telegram – 1 shilling under 100 miles, 1 shilling and sixpence under 200 miles, 2 shillings anything above.[51] As agreed, on the Electric/British & Irish handbills announcing the new prices the message was clear:

> The directors were therefore obliged to maintain a most anomalous Tariff, having been compelled, contrary to their convictions, to accept the shilling rate at the towns touched by their competitors and at the same time to maintain their old tariff at all the remaining places, an arrangement naturally most unsatisfactory to the general public. The uniform shilling tariff has had a fair trial, and failed; the Directors however believe that the entire revision of rates which they are now enabled to announce may be looked upon as permanent, for whilst the charges are calculated upon an extremely reasonable basis, it is hoped they will yield sufficient profit to justify their continuance.[52]

The price cartel, this time official and well advertised, unlike the one in 1855, decreed the success of the business strategy adopted by the main telegraph companies since the mid-1850s. Yet by late 1865, the telegraph sector presented the same problems that had emerged at the beginning of the decade: the national grid had not been extended, with the exception of the already existing circuits in the main industrial and commercial cities/towns; the technological stagnation continued; telegram prices had gone up again. In the very moment in which the cartel decreed the victory of the oligopoly, it also decreed the beginning of its end. The business world and the press had grown used to an economic flat charge and hostile as they were, set off a process which would shortly lead to the nationalization of the service.[53]

---

remunerative, the United Kingdom Company shall publicly in each town where they have a station declare by printed Bills or otherwise, that the 1/- rate has proved a failure", BT, *United Kingdom Electric Telegraph Company*, TGH 2/2/1, Letter from Electric to United, 7 June 1865; "My dear Sir, We...and Magnetic limit them at present to 10 words. Do you intend. To adopt our way, or what do you wish done, please consider and let me know", Ibid., "Letter from Electric to United", 23 June 1865.

[51] BT, *United Kingdom Electric Telegraph Company*, TGH 2/2/1, *The Electric and International Telegraph Company – General revision of tariff* [handbill], 3 July 1865.

[52] Ibid.

[53] Examples of some press comments: "the press of this country is at this moment in a very humiliating and undignified position. It claims, an justly claims, to be the corrector of abuses, the enemy of monopolies, and the pioneer of progress. Yet I cannot

5.3 Submarine telegraphy

The rapid, intense development of British submarine telegraphy between 1855 and 1866 was complementary to the consolidation of the landline oligopoly. While one part of the telegraph market on land tended to adopt business policies that aimed at low investment and a high redistribution of the profits, which attracted a great many small investors but caused a technological stagnation, submarine

---

correct its own abuses, is the victim of a threefold monopoly and lies helpless and prostrate before what is really a feeble obstruction. We allude, of course, to the pitiable attitude in which it has hitherto been content to remain towards the Telegraph Companies. Scarcely a day passes but the public read in some leading provincial newspaper doleful or angry complaints of the delays, blunders, and neglects in the transmission of news by the electric wires", *Dundee Advertiser*, 15 August 1865; "Referring to the recent announcement, that the uniform shilling rate for telegrams is to be abandoned, the *Northern Whig* says: This is exactly what we anticipated. These telegraph companies have now attained to such a position of monopoly that if some legislative check be not applied to their rapacity, the public will be literally at their mercy. The next Parliament, there is little reason to doubt, will have to regulate the powers of the telegraph companies, and place on them the same restrictions as are imposed on railway companies. The combination policy of the Magnetic Company has created a dangerous monopoly which some member of the new Parliament will do himself credit by restricting. The Magnetic Company have steadily bought up opposition, exacting from the parties so bought off; and it appears that the public are now to come under like pressure. Almost unobserved, a new power has been functions, and seems disposed to use its monopoly unscrupulously. The telegraph companies have quietly grasped the position of uncontrolled power which the railway companies sought to obtain, but which a wise legislation in the public interest took from them", "The Monopoly of Telegraph Companies", *Fife Herald*, 13 July 1865; "The proprietors of Scotch papers are in a rage with the Telegraph Companies, and though moved more immediately by their own interests, they are in this matter the spokesmen of the public. What with patents, royalties laws, privileges, private agreements among the Companies themselves, and old arrangements with the railways, the public has lost the greater part of the benefit which science at one time seemed to have bestowed. The telegraph is practically in the hands of a great monopoly which not only charges what it pleases and does the work when it pleases, but leaves whole districts untouched, arguing that perhaps they might not pay. In many places it is actually quicker to send a letter than a telegraphic message, in all the high rates of charge reduce the number and length of messages to the barest need of the senders, while the demand for off distances, are, except to the very rich simply propitiatory. ... There are we believe, two devices which would secure to the public the full use of the best means of communication yet discovered, and there are only two. The first, and the one we should prefer, is for the State to become the sole agent for telegraphs, to manage the lines as it manages the Post Office, to make them universal, leaving the great cities to pay for the little villages, counties like Lancashire for counties like Caithness, and to levy a single tariff, uniform for twenty words. ... The alternative in the matter of telegraphs is a system of real free trade, established by a general Act", "Telegraphic Free Trade", *Bradford Observer*, 31 August 1865.

telegraphy was based on great projects which would yield their fruits over the long period and consequently attracted big capital and technological innovations. The transfer of capital and technological know-how from landlines to submarine cables is the other side of the coin of the landlocked oligopoly.

In the early 1850s, the cables laid covered only brief distances and mainly served to complete the national network (the Ireland/ Scotland cable) or link up with neighbouring states (Channel cable). As we have seen, the tendency was carried on by landline companies, who between 1855 and 1860 continued investing in the sector and laid short cables to other European states (Holland, Germany and Denmark). Meanwhile, in 1855 more ambitious projects began to emerge, which foresaw longer cables that were able to link different continents. In chronological order, the first attempt was by John Brett, who between 1853 and 1857, with the Compagnie du Télégraphe Électrique Sous-marin de la Méditerranée pour la correspondance avec l'Algerie et les Indes, aimed at linking up Europe with Africa. Brett was to lay a first submarine cable between Italy and Corsica, a second between Corsica and Sardinia and a third one between Sardinia and Bona in Algeria. The system was to be completed by two landlines crossing Corsica and Sardinia.[54] By 1854 the company had completed its line from Italy to Sardinia, but the part between Sardinia and Algeria sank to the seabed, dragged down by its own weight and the precarious hold of machines used for its immersion.[55] The cable was laid on a second attempt in 1857[56] but because of the structure of the seabed and the specific conditions of the currents, it stopped working after only two years.[57] Though the company had its headquarters in Paris and received notable subsidies from

---

[54] Charles Bright, *Submarine Telegraphs. Their History, Construction, and Working*, London, Crosby Lockwood and Son, 1898, pp. 16–21.

[55] Pascal Griset, *Entreprise, technologie et souveraineté: les télécommunications transatlantiques de la France*, Paris, Editions Rive Droite, 1996, p. 44.

[56] "The attempts to lay down this line have failed on two previous occasions, and it is most gratifying to the promoters and to the public generally that the perseverance of Mr. Newall has at length been rewarded with success, The line now completed crosses the Mediterranean almost in direct line north and south from Spezzia [*sic.*] to Bona, on the Algerian coast", "Completion of the Mediterranean and progress of other submarine telegraphs", *Morning Post*, 2 November 1857.

[57] Ernesto D'Amico, *Cenni sull'amministrazione dei telegrafi in Italia dalle origini all'anno 1885*, Roma, Tipografia Cecchini, 1886, p. 80.

both the French and Piedmontese Governments,[58] it was mostly financed by British capital. Significantly, its board members included James Carmichael and Samuel Laing, two businessmen who were shareholders in the Submarine/European and the British and very close to John Brett.

While the cable to Algeria had attracted the attention of experts in the sector and the press,[59] the event that caught the public's imagination was the laying of the first transatlantic cable. In 1857 and 1858 there were three attempts to immerse cables, followed step by step by the British newspapers, which completely ignored the progressive consolidation of the oligopoly on land.[60] When in August 1858 the cable was finally in its place, the press went ballistic, with enthusiastic articles detailing the success of the undertaking, the official banquets and public celebrations.[61] The collective joy lasted about three months, until the cable broke down forever between September and October 1858.[62] At that point the press went on the attack and for months delved into the possible reasons for a failure that proved more astounding that the initial success.[63]

---

[58] Initially, Brett had drawn up contracts with both the French and Piedmontese administrations. In general terms, the conventions provided for the building, laying and maintenance of the cables, and the running of the service on the submarine lines with the total responsibility of the Compagnie du Télégraphe, which committed itself to take the line up to the African coast. The French government would have to pay £180,000 corresponding to 5% for 50 years while the Piedmontese government promised to pay 150,000 Piedmontese lire per annum corresponding to 5% on the capital for the same period. Furthermore in paying the equivalent of 16,000 lire, the Piedmontese Telegraph Administration took control of the Sardinian landlines except one, which remained with the company. Simone Fari, *Una Penisola in comunicazione. Il servizio telegrafico dall'Unità alla Grande Guerra*, Bari, Cacucci Editore, 2008, pp. 73–75.

[59] The Chronicles of the *Annales Télégraphiques*, from 1855 to 1859, describe in great detail the operations that were to complete the mixed line that from La Spezia was to cross Corsica and Sardinia to reach Algeria. See: Monsieur Ailhaud, *Pose du cable sous marin entre la Sardaigne et l'Algérie*, in "Annales Télégraphiques", 1858, pp. 209–219.

[60] Daniel Headrick, *The Invisible Weapon. Telecommunications and International Politics*, New York-London, Oxford University Press, 1991, pp. 17–18.

[61] Griset, *Entreprise*, p. 48. The following are some of the enthusiastic articles published between August and October 1858: "Mr Bright, the engineer of the Atlantic Telegraph Company", *Liverpool Daily Post*, 12 August 1858; "History of the Atlantic Telegraph", *Birmingham Daily Post*, 24 August 1858; "The Atlantic Wedding-ring", *Lloyd's Weekly Newspaper*, 5 September, 1858.

[62] "The Atlantic Cable: Suspension of the communication", *Bradford Observer*, 9 September 1858.

[63] Here, too, are a few examples of the numerous articles published on the subject: *London Daily News*, 21 September 1858; "The Scientific History of the Atlantic Cable", *Huddersfield Chronicle*, 25 September 1858.

Meanwhile, two distinct groups of entrepreneurs had presented two alternative telegraph links with India to the Government. The first, promoted by the Brett brothers with the establishment of the European and Indian Junction Telegraph Company, planned to connect the Mediterranean with the Persian Gulf and then proceed with a submarine cable.[64] The second, backed by the Gisborne brothers and their Red Sea and India Telegraph Company, proposed linking Egypt with Karachi via two submarine cables. The Government originally chose the first project, but had to abandon it because the Turks gave no authorization. The following year, against the background of the 1857 Indian Mutiny, the Government gave its backing to the Red Sea project, guaranteeing an even larger subsidy than had been given to the Atlantic.[65] The Red Sea proceeded with the immersion of two cables (Suez-Aden, Aden-Karachi) in 1859 and early 1860, but the first broke as it was being laid and the second stopped working after the first trial month.[66]

The double failure of the Atlantic and the Red Sea induced the Government to create a joint committee, with half of its members representatives of the Board of Trade and the other half delegates from the Atlantic. Blame for the debacle of the Atlantic cable was mostly attributed to Edward Whitehouse, who was said to have burnt the cable by using too strong an electric current.[67] In reality,

---

[64] "The prospectus of a telegraph company to complete the chain of electric communication between Europe and India has been published. The title of the company is the European and India Junction Telegraph Company", *London Standard*, 28 July 1856; "Notice is given that application will be made to Parliament next session for Acts incorporating respectively the European and Indian Junction Telegraph Company. Supposing these projects to be carried out, they will constitute two of the greatest enterprises of modern times. The former is to compose the intermediate link in the chain of telegraphic communication between London and India; while the latter company proposes to join the wires of the American telegraphs with those of Europe. What would our forefathers have thought of a project for enabling messages to be almost instantaneously transmitted from New Orleans to Newfoundland, thence under the Atlantic Ocean, and so across India to Calcutta? Yet it is quite within the bounds of possibility that within be consummated", "The Electric Telegraph between London and India", *Worcester Journal*, 22 November 1856.

[65] John Cell, *British Colonial Administration in the Mid 19th Century: The Policy Making Process*, New Haven, Yale University Press, 1970, pp. 226–233.

[66] Bright, *Submarine Cables*, pp. 57–58.

[67] Great Britain, Submarine Telegraph Committee, *Report of the Joint Committee appointed by the Lords of the Committee of privy Council for Trade and the Atlantic Telegraph Company to Inquire into the construction of submarine telegraph cables: together with the minutes of evidence and appendix*, London, 1861.

the promoters of the initiative were mostly to blame, because they had moved too quickly and made many little mistakes, which had caused structural damage to the cable, worsened by the extreme conditions on the seabed.[68] The commission found that apart from the mistakes made, the cable had managed to function for three months in 1858 and therefore work on the Atlantic could continue. It was just a case of waiting for the end of the American Civil War. Another cable was lowered in 1865 but got lost in the depths of the ocean. John Pender, who had taken over the company, found the finances for a new cable, which was laid without complications and started operating in 1866. A few weeks later, the 1865 cable was retrieved and activated, so decreeing the triumph of British telegraphy.[69]

The influence of land telegraphy on the development of submarine telegraphs between 1855 and 1866 is to be seen principally in the flow of capital and managers between the two sectors. We have already shown how the capitalization of the Atlantic was promoted directly by the Magnetic, which supplied a good nine of its directors. In addition, the Atlantic's secretary was Saward, who had previously held the same position in the British, while Charles Bright, who had made the Magnetic's cables, was the engineer responsible for laying the two transatlantic cables. In his turn, John Pender, a director of the Atlantic in 1856, who led the operation of capitalization of the company in 1866, had been a shareholder and director of the British and the British & Irish. It was he who in 1866 brought Brassey, an old-time shareholder and point of reference in the Electric into the Atlantic.[70] The same 1859–1860 enquiry, which gave the green light to continue the transatlantic cable, was composed mainly of men coming from land telegraphs. The representatives of the Board of Trade were Robert Stephenson, who died a few days after the beginning of the meetings, Charles Wheatstone, George Bidder and Douglas Galton, from the War Office, and William Fairbain, chairman of the British Association. The delegates of the Atlantic were Edwin and Latimer Clark, both ex-chiefs of the Electric, Cromwell F. Varley, a

---

[68] Donard de Cogan, "Dr. E.O.W. Whitehouse and the 1858 Trans-Atlantic Cable", *History of Technology*, 10, 1985, pp. 1–15.

[69] Vary T. Coates, Bernard Finn, *A Retrospective Technology Assessment: Submarine Telegraphy. The Transatlantic Cables of 1866*, San Francisco, San Francisco Press, 1979.

[70] Bright, *Submarine Telegraphs*, p. 82.

former electrician in the same company and the already mentioned George Saward.[71]

Although at the beginning the technical and scientific background of submarine telegraphy was more or less the same as that of the land telegraphs, as is seen by the passage of engineers, electricians and managers from the latter to the former, it presented some new technical problems, which could only be solved in the period 1855–1865, thanks to the continual technological innovations.

Laying the cables under seas encountered at least four grades of difficulty: constructing impermeable cables capable of transmitting electric current; sounding the seabed to ensure the safe laying of the cable; transport and immersion of the cable; reception of electronic impulses over long distances.[72] The first problem, of the cable's impermeability was mostly resolved with the application of gutta-percha.[73] Submarine cables were already being constructed in the early 1850s[74] as the years passed, they were gradually perfected while their structure remained more or less the same.[75] There were one or two conducting wires in the centre,[76] which served for transmitting the electric current, and the wires were then wrapped in a layer of gutta-percha to ensure they were impermeable.[77] The insulated wires were held together with a mixture of tar and hemp, again to increase impermeability. To ensure resistance to possible external mechanical agents, the cable was armoured, i.e., strengthened on

---

[71] Ibid., p. 59.

[72] Carlo Matteucci, *Manuale di telegrafia elettrica*, Torino, 1861, p. 384.

[73] William Hooper, *Indian Rubber considered in reference to its Applicability as an insulator for telegraphic conductors*, read at the Birmingham Meeting of the British Association in Section A, September 7th, 1865, and again following day, in section G, by request of the Committee, stored in K I, British Telecom archives, London.

[74] Robert S. Newall, *Facts and Observations relating to the invention of the submarine cable and the manufacture and laying of the first cable between Dover and Calais in 1851*, London, E. & F. SPON, 1882.

[75] About stagnation in submarine cable technology, see: Bernard S. Finn, "Submarine Telegraphy: A Study in Technical Stagnation", in Bernard S. Finn, Daqing Yang (eds), *Communication under the Seas: The Evolving Cable Network and Its Implication*, Cambridge, MA, MIT Press, 2009, pp. 9–24.

[76] Right from the beginning, it was deemed opportune to insert more than one conducting wire inside the cable, in case one got damaged.

[77] T. Seeligmann, *Indian Rubber and Gutta Percha: A Complete Practical Treatise on Indian Rubber and Gutta Percha in Their Historical, Botanical, Arboricultural, Mechanical, Chemical and Electrical Aspects*, London, Scott Greenwood, 1903.

the outside by a spiral of thick steel or iron wires.[78] The typology of the seabed had to be examined before the cable could be laid and care taken to avoid at all costs, for example, suspending the cable over two peaks, so that the middle part floated at the mercy of the currents. Laying transatlantic cables was only made possible by the notable study of oceanography,[79] well exemplified by the work of Matthew Maury.[80]

Given the steel armour, the cables were very heavy. The kind of ships needed had to be able to transport great weights and be equipped with mechanisms for laying cables. At first, very large craft adapted for transporting and laying cables were used. Then new ships were custom built. They were equipped with enormous rollers that the cables were wrapped round, while enormous chain-pulleys connected to the rollers slowly lowered the cable into the water. During the operation the personnel had to be careful that the cable was not damaged, which would have ruined all the work done. To avoid any ugly surprises, the cable was tested as it went down. Sometimes the chain pulleys could not hold the weight and dropped the cable into water, totally out of control.[81] In order to receive the weak electric current that had crossed thousands of miles in an inter-continental cable, telegraphs more sensitive than the Morse, like the syphon, were invented.[82]

To sum up, the technology required for building, laying and running cables was complex, refined and called for heavy invest-ment, both in the infrastructures and in research and development.

---

[78] The outside armour could favour the dispersal of electric energy via "inductive currents". The problem was resolved with different expedients by many scientists including Wheatstone and Siemens. See: Werner Siemens, *Apparati per l'esercizio di lunghe linee telegrafiche sottomarine di Werner Siemens e Halske*, Firenze, Tipografia Bencini, 1867.

[79] John Steel Gordon. *A Thread across the Ocean. The Heroic Story of the Transatlantic Cable*, New York, Perennial, 2002, p. 37.

[80] Frances Leigh Williams, *Matthew Fontaine Maury. Scientist of the Sea*, New Jersey, Rutgers University Press, 1963.

[81] Matteucci, *Manuale di telegrafia elettrica*, pp. 391–394.

[82] The syphon was a very simple mechanism that was limited to amplifying very weak signals. See: Ludovic A. Ternant, *Transmission des signaux par les cables sous-marins*, Paris, Ducher, 1875; Ludovic A. Ternant, *Le siphon-recorder et le curb-sender automatique*, Paris, G. Masson, 1882.

5.4   Turning points and constitutive choices

The first 20 years of British telegraphs had laid the conditions for a limited market, available to a restricted number of companies closely linked to the big capital of the railway and financial worlds. Nevertheless, exogenous and endogenous factors began to appear about 1855, i.e., the increase of small shareholders in the capital of the telegraph companies and the deterioration of the gutta-percha in the underground cables.

The arrival of the small shareholders, incentivized by the telegraph boom of the previous years, led to the adoption of business strategies aiming at immediate benefits and the redistribution of the profits, and favouring cartel agreements to ensure the conservation of the status quo.

Nevertheless, the turning point was probably determined by the deterioration of the gutta-percha, an exogenous happening, in some ways unforeseeable. The Magnetic had invested capital, especially human, in bypassing its rivals on a technological level with a network of underground lines. Its collapse encouraged the immediate transfer of technicians and technologies from land to submarine telegraphy. Faced by the impossible task of implementing the ruined underground network, many engineers and managers exploited the knowledge acquired in matters of insulating electrical cables with gutta-percha and recycled themselves in the sector of submarine telegraphs, which promised big profits.

The Government, which had stimulated investments in gutta-percha by obliging cable to be laid underground in urban areas and financed transcontinental undertakings, never showed much interest in land telegraphy. The sudden and intense boom in submarine telegraphy accelerated the consolidation of an oligopoly, whose foundations were laid right at the beginnings of telegraphy. Paradoxically, however, the triumph of the oligopoly in 1865 was the root cause of a drastic change of regime in the telecommunications market, which in its turn would be stimulated by the continued development of the international submarine network.

6
Nationalization

Twenty two years ago, a few far seeing men ... originated the telegraphy of the whole world. They were called visionary enthusiasts and madmen and all sorts of hard names. They were not supported by the public; they were not assisted by the Government; they were not able to pay dividends, and if they had lost the whole of their capital they certainly never would have been reimbursed either by the public or by the Government, and it is only after years of patient and prudent management and when the concern has become an established success that the Government proposes to acquire it. ... it is the game that the big boy plays with the little boy in the street 'Heads I win, tails you lose'.[1]

1865 marked the second turning point in the telegraph communications market with the abolition of the flat charge and public opinion building up in favour of nationalization as the only way to redress an inefficient and costly telegraph service. The following year Frank Ives Scudamore drew up for the Post Office a first official plan for the Government's acquisition of British telegraph network. Then in early 1868 the Bill was presented in Parliament and approved with only a few minor amendments. The next year a second Telegraph Bill was discussed, with the aim of establishing a government monopoly and the indemnity for the telegraph companies. As of 1 January 1870

---

[1] BT, *Electric and International Telegraph Company*, TGE 1/2, *Half Yearly Meetings*, February 1868.

the British network was Government-run, as in all European states. The surprising speed with which the project was realized was due basically to two facts. Firstly, customer dissatisfaction was by then well entrenched and rising rapidly with the strengthening of the oligopoly. Secondly, the opposition of the telegraph companies and the railroads gave way at once, as they were all open to financial negotiations.

Apparently, nationalization was guided by the economic principles underlying the concept of a natural monopoly upheld by greater economies of scale, lowered tariffs, universalized service and a high economic standardization. But here again, long-term features are essential for any understanding of the phenomenon.

6.1   Plans and inquiries

Up to 1865 a Government management had never been publicly taken into consideration as a plausible solution to the inefficiency of the telegraphic service.[2] In the early 1850s the press and businessmen had backed the entry into the market of the British, the Magnetic and any other rivals, convinced that free trade would do away with the aberrations of the monopoly. In the same way, in the early 1860s they had promoted the entry of the United, and its tariff plan as the only remedy to the oligopoly.

After the events in July 1865, more radical moves were made. When the first protests were heard, George Harrison, leader of the Edinburgh Chamber of Commerce put forward the idea of placing the telegraph service in the hands of the Government,[3] and a

---

[2] The only exception was Andrew Wynter, "The Electric Telegraph", *Quarterly Review*, 95, June, 1854, pp. 150–151. Nevertheless, there was no follow-up to this proposal; it remained an isolated case.

[3] "That was a specimen of how telegraphic affairs were managed. He had long been of opinion that in some kinds of business a regulated monopoly was better than an apparent competition. In many things competition seemed impossible. They knew very well that although the Post-office was originated by private enterprise, it would have been impossible to have given the same facilities in this important department of public service had the working of it been left to public competition. The (sic.) also had experience of railway competition and had found it necessary to enter into negotiations to allow amalgamations, having found out that a regulated amalgamation was in some measure better than a sham competition. They were aware the Government had appointed a royal commission to see whether it was possible to go further into

committee was appointed to investigate the conditions of the telegraph in Great Britain with the idea of suggesting improvements.[4] In October the committee pinpointed at least three main defects in the private companies. It complained of the excessive tariffs, underlined an excessive number of mistakes and delays and pointed out that the number of cities covered by the telegraph service (about 1,000) was far lower that the number with post offices (more than 10,000).[5] The committee then proposed three solutions for an improvement. The first was the creation of a private monopoly under Government supervision, which would keep its tariffs and dividends under a ceiling established by the law. The second proposal spoke for free

the matter of railways. The difficulties of doing so were, of course, something enormous; but there would be very much less difficulty in this comparatively early stage of telegraphic communication if the government were to see fit to take charge of that department of the public service. Although he was not disposed to make any decided representation regarding the desirableness of such a step being taken, he thought at the same time it was one very well worthy of consideration", "The Monopoly of the Telegraph Companies", *The Waterford Mail*, 11 August 1865.

[4] "What he proposed was to ask the Chamber to appoint a committee to consider the state of telegraphic communication, to get information on the subject from public offices, newspaper offices, and private individuals, and to bring up a report showing how the present system works, and whether they were in the position to recommend something that might be useful to the public. The motion was unanimously taken", Ibid.

[5] "The chief grounds of complaint against the present telegraphic system are: 1st. The high charges for the transmission of messages. 2nd. The frequent and vexatious delays in their delivery, and their inaccurate rendering. 3rd. That many important towns, and even whole districts, are unsupplied with telegraphic communication.... 1st. As to charges. The present tariff of all three companies carrying on business is in Great Britain 1s. per twenty words for distances of 100 miles and under; 1s. 6d. from 100 to 200 miles; and 2s if above 200 miles. The wires of the United Kingdom Company do not extend to Ireland, and the charges made by the two companies which have lines to that country are at present 3s. from Edinburgh to Dublin, Belfast, and the principal towns on the east coast; 4s. to nearly all other places; and 5s to Valentia.... As to delay and inaccuracy in delivery of messages. The evidence which your company has received as to the delays in the delivery of messages is so overwhelming, that they are compelled to believe that the system is in this respect capable of great improvement. Of their frequent incorrectness, and the serious consequences which have sometimes resulted from most unwarrantable blunders almost every person using the telegraph has had reason to complain. 3rd. As to the deficiency in the means of telegraphic communication. Your committee need only say that the three existing companies provide means of communication between about 1000 separate places, while the General Post-Office has in England 544 head offices and 7250 sub-offices; in Scotland 114 head and 1180 sub-offices; and in Ireland 138 head and 1459 sub-offices – making in all 10,685. We may add that there is no telegraphic communication to or from such places in Scotland as St. Andrews, Leven, Thurso, or Wick", "Edinburgh Chamber of Commerce", *Caledonian Mercury*, 13 October 1865.

trade, facilitated by the abolition of all regulations on constructing telegraph lines and Board of Trade control over the market. The solution favoured by the committee was the third, a Government takeover of the network, to be run by the Post Office, which the committee felt would bring about automatically the reintroduction of the flat charge.[6]

The radical choice of nationalization was the result of two trends already in the air in earlier decades. On one hand, Liberal economists had gradually become more tolerant of Government-run communications as a presupposition for free trade. For example, Adam Smith held already that the post was the only mercantile project that could be happily run by any Government because of the high fixed costs and the positive externalities coming from it.[7] Jeremy Bentham was

---

[6] "The Edinburgh Chamber of Commerce do not recommend any specific plan by which the present system of telegraphic communication can be improved; but suggest three important reforms. The first is the amalgamation of the three existing companies, with a moderate tariff, to be further reduced as the profits grow beyond a certain limit; but this could be effected only by the consent of the parties. The second project is more feasible. It is to adopt a complete system of free trade, by passing a general act authorising any person or association of persons to erect telegraph posts along any public way throughout the United Kingdom. We would suggest that a clause be added compelling the railway companies to give the use of their lines to all that applied for them for telegraphic purposes on payment of the same amount of rent as they might be receiving from any other company. The Edinburgh people seem to have a leaning towards the purchase of wires by the government, which we certainly do not share. There is already on the statute-book an act authorising the State to acquire the railways, and the objections to such a measure have been repeatedly pointed out. There is less to be said against the transfer of the telegraphs to the State, for the capital involved is comparatively small – somewhere under £ 2,500,000 – and the influence thereby conferred on the government would not be so dangerous. We have no objection to see the question fully argued out; but our feeling is decidedly against imposing on the State any duty which can reasonably be left to private enterprise", "The Necessity of Improved Telegraphic Communication", *Dublin Evening Mail*, 18 October 1865. For the complete text of the report, see BPP, 1867–1868, Vol. 41, *Report of committee appointed by the Chamber of Commerce to consider the present condition of Telegraphic Communication throughout the United Kingdom.*

[7] "The Post-office is properly a mercantile project. The government advances the expense of establishing the different offices, and of buying or hiring the necessary horses or carriages, and is repaid with large profit by the duties upon what is carried. It is perhaps the only mercantile project which has been successfully managed by, I believe, every sort of Government. The capital to be advanced is not very considerable. There is no mystery in the business. The returns are not only certain, but immediate. The exact, successful, and profitable management of a Post-office by the Government, is due to the fact that the public being interested to the same extent with the Government in the proper administration of the office, can act, and will act, with great vigilance on the Post-Office officials. The motive of the Government is the enlargement of its revenue

convinced, however, that Government run communications were a precondition for maintaining the freedom of the citizens.[8] Lastly, shortly before nationalization W.S. Jevons confirmed the same line of thought and stated that the Post possessed all the requisites for being run directly by the state, and a condition that could be extended to the telegraphs.[9] In short, if the telegraphs were nationalized, the operation would be legitimized by the dominant political economy.

---

from the Post; that of the public is the regular, frequent, and punctual distribution of its correspondence. There is and has been only one difference between the Government and the public as regards this branch of administrative business. The Government is afraid of cheapening the means of transit, lest the revenue should suffer. The public insists that a public service should not be considered as a mere means of making a revenue, for in this case legitimate profit becomes sheer taxation, and taxation of an invidious and differential kind. Besides it argues that cheap communication is likely to swell the gross profits of the Post-office, by the increased use of the convenience. And the public is in the right", Adam Smith, *An Inquiry into the Nature and Causes of the Wealth of Nations*, Oxford, Clarendon Press, 1880, pp. 405–406.

[8] "By instructions, excitation and faculty of correspondence – by these three instruments in conjunction, and not by any one or two them alone – can the national mind be kept in a state of appropriate preparation – a state of preparation for eventual resistance. It is by the conjunct application of all these instruments, that minds are put and kept in a proper state of discipline, as bodies are by the military exercise. From this state of full and constant preparation, result two perfectly distinct, though so intimately connected uses: 1. Effecting a change in government, if ever, and when necessary; 2. In the meantime, preventing, or at least retarding, the necessity, by the constant application of a check to misrule as applied to individual cases – to misrule in all its several shapes. Necessary to instruction – to excitation – in a word, to a state of preparation directed to this purpose, – is (who does not see it?) the perfectly unrestrained communication of ideas on every subject within the field of government; the communication, by vehicles of all sorts – by signs of all sorts; signs to the ear – signs to the eye – by spoken language – by written, including printed language – by the liberty of the tongue, by the liberty of the writing-desk, by the liberty of the *Post-Office* – by the liberty of the *press*. The characteristic, then, of *an undespotic* government – in a word, of every government that has any tenable claim to the appellation of a *good government* – is, the allowing, and giving facility to this communication; and this not only for instruction, but for excitation – not only for instruction and excitation, but also for correspondence; and this, again, for the purpose of affording and keeping on foot every facility for eventual resistance – for resistance to government, and thence, should necessity require, for a change in government". Jeremy Bentham, "On the Liberty of the Press, and Public Discussion, Addressed to the Spanish People", in John Bowring (ed.), *The Works of Jeremy Bentham*, Part VII, Edinburgh, William Tait, 1839.

[9] "It seems to me that the State management possesses advantages under the following conditions: 1. Where numberless wide-spread operations can only be efficiently connected, united and coordinated, in a single, all extensive Government system. 2. Where the operations possess an invariable routine-like character. 3. Where they are performed under the public eye or for the service of individuals, who will immediately detect and expose any failure or laxity. 4. Where there is but little capital expenditure, so that each year's revenue and expense account shall represent, with sufficient

On the other hand, in the mid-1850s some detailed projects for the nationalization of the British telegraphs were already being drawn up but were not widely circulated. The first two were drafted in 1854 and 1856, the first by Thomas Allan and the second by Frederick Ebenezer Baines, a Post Office official at the time, who between 1848 and 1855 had worked as a telegraph operator for the Electric.[10] In particular, Allan, whose attempt to incorporate United had failed, re-proposed as a Government monopoly the introduction of a new technology which would have allowed a flat charge. Opportunistically Allan hoped to sell to the Government what he had not managed to sell to privates.[11] Baines's project, presented

---

accuracy, the real commercial conditions of the department. It is apparent that all these conditions are combined in the highest perfection in the Post Office. It is a vast co-ordinated system, such as no private capitalists could maintain, unless, indeed, they were in undisputed possession of the field by virtue of a Government monopoly. The forwarding of letters is a purely routine and equable operation. Not a letter can be mislaid but someone will become aware of it, and by the published tables of mail departures and arrivals the public is enabled accurately to check the performance of the system. Its capital expenditure, too, is insignificant compared with its current expenditure. ... A telegraph system appears to me to possess the characteristics which favour unity and State management almost in as high a degree as the Post Office. If this be so, great advantages will undoubtedly be attained by the purchase of the telegraphs and their union, under the direction of the Post Office department", William Stanley Jevons, "On the Analogy between the Post Office, Telegraphs, and Other Systems of Conveyance of the United Kingdom, as Regards Government Control", in William Stanley Jevons, *Methods of Social Reforms and Other Papers*, London, MacMillan and Co., 1883, pp. 279–280.

[10] "In 1846, on the incorporation of the Electric Telegraph Company, I had acquired a taste for telegraphy, and at only fourteen years of age had mastered its principles and language; and when in April 1848, the Company and the Postmaster-General had made their agreement for bringing the "electrical Telegraph" to the Post Office, the company, at the suggestion of Mr. Rowland Hill and his brother, Mr. Frederic Hill (my uncle by marriage), took me into their service. ... Meanwhile, with the Electric and International Company I spent several years until, in 1855, the good friends who had helped me back to the Post Office, not as the official of the Telegraph Company, but as an established officer of the Postmaster-General", F.E. Baines, *Forty Years at the Post Office. A Personal Narrative*, London, Richard Bentley and Sons, 1895, Vol. I, pp. 143–159.

[11] "It may be assumed that a general and more extended use of the electric telegraph will follow the same general laws as the working of the penny post and cheap literature, and, from the wonderful development of these principles in the instance of the penny postage, the hypothesis became reduced to an anticipated fact.... Hitherto, in this country, there has been no fixed principle adopted in telegraphing, arising from the want of system and experience, in the first instance, great costs in the construction of works, and fragmentary or isolated arrangements; yet wherever a reduction of charge has been made, a marvellously corresponding increase in quantity [of telegrams] has been the result", BPP, 1867–1868, Vol. 41, *Thomas Allan's Reasons for the Government Annexing an Electric Telegraph System to the General Post Office*, p. 40.

to the Treasury with the authorization of the Postmaster General, seemed to follow Allan's guidelines: a Government monopoly, an extended service and the introduction of a flat charge.[12] Its differences lay in the fact that it was extremely detailed, with its proposal backed by a lot of statistics.[13] Also, Baines was convinced he had the moral duty to universalize a service which up to then had been exclusive.[14]

At the beginning of the 1860s, just when the problems connected with the oligopoly started to emerge, Ricardo and Burchell, a railway man and minor shareholder of the Electric since 1851,[15] sent a memorandum in favour of nationalization to Gladstone, the Chancellor of Exchequer. The two promoters were convinced that a public management would be more efficient because its managers could quietly run the service in the interests of the community, which

---

[12] "The main features of my scheme were: (a) A Government system of electric telegraphs, with privilege of exclusive transmission of public telegrams, similar to that enjoyed by the Post-office in respect of letters; in short, a monopoly. (b) The incorporation of the proposed system with the Post Office, and extension of the wires in the first instance to the post-office of every post-town in England, Wales and Scotland.... (c) The adoption of a uniform charge of 6d. for each message of twenty words between any two post-towns, inclusive of delivery within the limits of the terminal town", Baines, *Forty Years*, p. 300.

[13] "(1) The Government departments would be placed in possession of a reliable means of instantaneous communication with all the out-ports and inland towns, as well as with the dock-yards and arsenals of Portsmouth, Plymouth, Pembroke, Sheerness, Woolwich, Chatham, and that the same advantage would be afforded to the public at a reasonable charge. (2) Central telegraph offices would be opened at 470 post-towns which previously had no telegraphic communication whatever; and at 210 post-towns there would be central Post-Office telegraphs instead of railway-station telegraphs; so that telegraph offices; and there was to be eventually an extension of the wires to a part of the 8,000 or 9,000 sub-offices, if not at all. (3) Remote districts of Great Britain would be brought into immediate communication with the centres of commerce; their trade would be facilitated, and their progress assisted. (4) A complete separation of the railway telegraphic service from that of the public, which was strongly advocated, would tend to the safety of passengers travelling by railway, and the rapid transmission of public despatches. (5) The consolidation of the system under one management would, it was clear, remove the liability to error resulting from the repetition of despatches at the termini of distinct telegraph companies' lines. (6) Lastly, a progressive accession to the net public revenue, rising from a minimum of upwards of £ 50,000 a year, could be, not without reason, anticipated", BPP, 1867–1868, Vol. 41, *F.E. Baines to Treasury, 1 March 1856*, p. 42.

[14] "[The high rates were] sufficient to place the telegraph within the reach of only the more opulent classes, whereas it is of the highest importance that the facilities afforded by its use should be readily available to all classes", Ibid.

[15] BT, *Electric Telegraph Company*, TGA 1/6, *Shareholder Books 1846–1856*.

meant yet again low tariffs and extending the service without worrying about shareholder pressure. The project calculated the cost of buying the telegraph lines at two million sterling, and the possible annual profit at £60,000.[16]

This memorandum takes on notable interest if we consider that Ricardo was not only the main artificer of the expansion of the Electric in 1850 but also a strenuous defender of free trade in the Commons. Perry offers a fairly convincing explanation of Ricardo's behaviour, holding that Ricardo saw nationalization as an institutional reform which could favour free trade and eliminate the inefficiencies of the oligopoly.[17] It was similar to the position taken by Richard Cobden, the chief spokesman of the Anti-Corn Law League during various discussions on the introduction of the Penny Post.[18] Therefore, according to Perry, Ricardo's memorandum is perfectly in line with his ideas on free trade, and not contradictory, as it could appear.[19] The reasoning in the memorandum is also linked to Ricardo's experience as a manager. He was exited from the chairmanship and then from the board of the Electric because his strategy of extending and defending the dominant position of the Electric was on a collision course with the interests of a broad-based shareholding calling for an ample redistribution of the profits. In the same way Ricardo had tried to avoid cartel agreements because he was convinced that the oligopoly would generate a series of serious problems over time, including excessive tariffs. He had therefore reached the conclusion that free trade in the telegraphs market set off a vicious circle which went on strengthening an oligopolistic format, via the expansion of the service and the consequent increase in the number of small shareholders. Therefore, the only possible solution was to put the telegraph in the expert hands of the Post Office, which ever since the introduction of the Penny Post had proved capable of

---

[16] BPP, 1867–1868, Vol. 41, *Memorandum in Support of the Expediency of the Telegraphic Communication being Placed in the Hands of Her Majesty's Government.*

[17] Charles R. Perry, *The Victorian Post Office. The Growth of a Bureaucracy*, Woolbridge, The Royal Historical Society-The Boydell Press, 1992.

[18] Nicholas C. Edsall, *Richard Cobden Independent Radical*, Cambridge, MA, Harvard University Press, 1986.

[19] Perry's argument is an answer to the doubts raised about the question in Jeffrey Kieve, *The Electric Telegraph, A Social and Economic History*, Devon, David and Charles Newton Abbot, 1973, pp. 120–121.

carrying out a profit-making and efficient public service aiming at universalization.

Lord Stanley, Postmaster General at the time, had already for many years held that his department was capable of running the telegraph service better than the private companies, an opinion corroborated by the reports. Nevertheless, it was only in September 1865, after the open agitation of the press that Lord Stanley decided to officially instruct his right-hand man, Frank Ives Scudamore, to draw up a detailed report on the private running of the telegraph system and the possibility of a Post Office takeover.[20] Scudamore, who in that period held the second most important position in the Post Office, and had to his credit great successes like the network of Post Office saving banks, the introduction of the postcard, and life insurance, threw himself wholly into the undertaking.[21] Some years later he confirmed officially that he had received the mandate in September 1865, before the findings of the Edinburgh Chamber of Commerce committee had been made public – a move to let it be understood that the idea of nationalization had come autonomously from the Post Office and that it did not act in answer to external pressure.[22]

But in August, before the committee was appointed, Harrison had already put forward in public the idea of a Government takeover, and when Lord Stanley sent for Scudamore, the campaign of the press and chambers of commerce against the oligopoly had been going strong for two months. Unlike Allan, Ricardo/Burchell and Baines, Scudamore had no working or managerial experience in telegraphs and had no knowledge of its technical and technological aspects, so he immersed himself totally in its workings and counted on his administrative skills. As he had previously shown, he was strongly in favour of the functional expansion of the Post Office, which he saw as a kind of multi-service company capable of generating notable returns and guaranteeing access to the citizenry of services which were difficult to acquire in a free market. In July 1866 he presented his report to Lord Stanley, who unsurprisingly reached the same conclusion as the Edinburgh committee, the tariffs were too high,

---

[20] Perry, *The Victorian Post Office*, pp. 94–95.
[21] Charles R. Perry, "Frank Ives Scudamore and the Post Office Telegraphs", *Albion: A Quarterly Journal Concerned with British Studies*, 12/4, 1980, pp. 350–367.
[22] Perry, *The Victorian Post Office*, p. 94.

transmissions often inaccurate and delivered with great delays, and above all, many towns and whole districts were not part of the network. Scudamore had also carried out a survey of 475 towns with a population of over 2,000, arriving at the conclusion that 30% were well served by the private lines, 40% had a telegraph office less than half a mile from the post office, 12% had one at a distance between ½ mile and a mile, and 18% either had no telegraph office or one at over a mile away. Like the Edinburgh committee, he considered three possible solutions (free trade, private monopoly and Government monopoly), though he was clearly in favour of the Post Office acquiring the telegraphs directly. He emphasized the fact that the Post Office would be able to run the service by opening telegraph offices in all towns with at least 2,000 inhabitants and reducing the tariff to one shilling for 20 words.[23] Like Baines, Scudamore was convinced that the telegraph service should be universalized and that the Post Office could do it, just as it had with the Penny Post.[24] Along the same lines as Ricardo and Burchell, he calculated that the entire operation would not cost more than £2.4m, an expenditure that would be remunerated by the profits coming from the increase in telegrams sent. Still on the same lines as Ricardo, Scudamore held that a public service would prove efficient because it underwent no shareholder pressure.[25]

Lord Stanley sent the report on to the Treasury shortly after receiving it. Only a few months later, in December 1866, the newspapers reported that the Post Office had drawn up a project for the purchase of the telegraph lines and would send it on to the relevant ministry.[26] However, the crisis over electoral reform toppled the

---

[23] BPP, 1867–1868, Vol. 41, *A Report to the Postmaster General upon Certain Proposals which have been made for Transferring to the Post Office the Control and Management of the Electric Telegraphs Throughout the United Kingdom.*

[24] "We have, in short in the telegraphic system of the United Kingdom precisely what we had in the postal system...before 1840, when the receptacles for letters were few in number, when the charge for transmission was excessive, and when the limits of free deliveries were so narrow that large numbers of letters were subjected to additional taxation before they reached the hands of the addressees", Ibid.

[25] "Little or no improvement can be expected so long as the working of the telegraphs is conducted in a wasteful competition with each other.... [The directors] have of necessity thought rather of the interest of their stockholders than of the interests of the whole community", Ibid.

[26] "A few days previously to the late Government leaving office, the draft of a bill providing for the purchase and working by Government of the telegraphic lines of

Liberal Government, which was replaced by the Conservatives, and the continued debate over extending franchise took up most of 1867. While Parliament's attention was directed elsewhere, Scudamore exploited the moment to carry out a powerful lobbying in order to conquer public opinion. He received at once the support of both leaders of the two main political parties. Gladstone, who had already collected the fruits of the creation of the Post Office saving banks, had great faith in him while Disraeli, who had recently obtained a parliamentary majority, wanted to take over the paternity of a project that was becoming more and more popular[27] thanks to the tireless public relations activity carried out by Scudamore, almost unconditionally backed by the press. The campaign reached its highest point in the spring of 1867, when Edwin Chadwick,[28] the social reformer,

---

the kingdom was in its possession. The scheme was the result of much previous correspondence with the officials of the leading telegraphic companies and with foreign Governments. It had been designed to submit a bill on the subject to Parliament had the Ministry continued to hold office. The ground on which the scheme would have been urged would have been the broad one of national advantage, in the adoption of rates far below those at present in force, and generally in the more systematic and efficient arrangements to be formed. The measure is understood to have received the attention of the present Administration, and they will probably bring forward a bill for that purpose. In any case, however, the scheme is certain to rank amongst the earliest measures of any other Ministry", "The Telegraphs and the Post-Office (from Observer)", *Morning Post*, 10 December 1866.

[27] Perry, *The Victorian Post Office*, pp. 100–101.

[28] "Let me recapitulate the chief conclusions which I have to submit. They are – that cheap and complete telegraphic communication, with the speedy and punctual delivery of telegrams, next to the cheap, speedy and punctual delivery of letters, is of the highest importance to the manufacturing, commercial and agricultural service of the country, as well as to the service of the state, for the army, the navy, and the police. That the present telegraphic communication is made unnecessarily dear, by the charges of several incomplete establishments, to perform a service that might be better performed by one. That there are in the United Kingdom ten thousands post offices, and a service of twenty thousand persons engaged in the collection and delivery of letters, available for the collection and delivery of telegrams... That, by the use of the existing postal establishments pervading all parts of the country, the postal service may – as is done by the postal establishments of foreign states – convey telegraphic messages at low rates of charge at a profit, which private companies can only convey at the same low rates, at a loss. ... That out of the charges of the separate establishments of separate private companies produced by consolidation, together with the increased traffic obtainable by telegraphic communication through the post at reduced tariffs, fair compensation may be made to the trading interests in public telegraphy which have been allowed to be formed, and that the proper responsible duties of the Government for the maintenance of the safe and free use of all public means of communication as a service, may be advantageously resumed without direct expense to the revenue, and with large

famous for promoting the Poor Law and sanitation improvement,[29] publicly defended it during a conference held by the Royal Society of Arts, which Scudamore attended,[30] as Jevons did during a meeting of the Manchester Statistical Society.[31] Only later, in November 1867, did the Government make official its intention to go ahead and seek parliamentary authorization to purchase the telegraph network.[32] Some months later, in February 1868, Scudamore presented an updated and

---

indirect advantage to it by the augmentation of the commerce and the production of the country", Edwin Chadwick, "On the Economy of Telegraphy as Part of a Public System of Postal Communication", *Journal of the Society of Arts*, 15/745, 1 March 1867, pp. 222–236.

[29] Samuel Edward Finer, *The Life and Times of Sir Edwin Chadwick*, London, Methuen, 1952; Anthony Brundage, *England's "Prussian Minister": Edwin Chadwick and the Politics of Government Growth, 1832–1854*, University Park, Pennsylvania State University Press, 1988.

[30] "The idea of associating telegraphic communication with the postal service is one which gains many advocates, and seems likely to command an increasing amount of public attention, in a recent discussion before the Society of Arts Mr. Scudamore stated that some time ago the officials of the Post Office proposed a plan, now under the consideration of the Government, by which the benefits of electric telegraphy would be immensely extended. Into the details of that proposal Mr. Scudamore did not feel at liberty to enter; but he went so far as to say, that he should be disappointed, if the adoption of the plan failed to place telegraphic facilities at the command of towns even with populations considerably less than 5,000", *London Standard*, 11 March 1867.

[31] "It is obvious in the first place that the public will be able and in fact obliged, constantly to test the efficiency of the proposed Government telegraphs, as they now test the efficiency of the Post Office. ... It is hardly necessary to point out, in the second place that a single Government telegraph system will possess great advantages from its unity, economy, and comprehensive character. Instead of two or three companies with parallel conterminous wires, and different sets of costly city stations, we shall have a single set of stations; and the very same wires, when aggregated into one body, will admit of more convenient arrangements, and more economical employment. ... Furthermore, great advantages will arise from an intimate connection between the telegraphs and the Post Office. In the country districts the telegraph office can readily be placed in the Post Office, and the postmaster can, for a moderate remuneration, be induced to act as telegraph clerk, just as small railway stations serve as telegraphic offices at present, the station-master or clerk being the operator. ... The Government, in short, could profitably extend its wires where any one of several competing companies would not be induced to go", Jevons, "On the Analogy", pp. 281–282. The paper was read at the Manchester Statistical Society, 10 April 1867.

[32] "The *Times* believes that the government, at the meeting on the 12th inst., decided on entering into arrangements with the principal telegraphic companies of the United Kingdom, with a view of placing the various lines under the management of the Post Office department. The details of the plan will be under the superintendence of the Post Office department", "The Telegraph and the Post Office", *Dublin Evening Mail*, 15 November 1867; "We perceived by the Parliamentary notices that the Government have now decided upon carrying out the plan, and that a bill will be introduced in the next session for the purpose", "Government and the Telegraphs", *The Glasgow Daily Herald*, 18 November 1867; "The Tory Government is apparently about to do a very

more detailed version of his report,[33] which became the basis of the Telegraph Bill presented in Parliament in April.

6.2   Against nationalization

As we have seen, the two great lobbies backing nationalization were the press and the chambers of commerce, both calling for a cheaper and more efficient service. The press, especially the Scottish and Irish newspapers, added criticism on the distribution of information, accusing private companies of controlling and at times manipulating news items.[34] The two lobbies acted in tandem. The press constructed its own long-term propaganda, expressing dissatisfaction going back to the 1850s, while the chambers of commerce gave further impulse to telegraph reform with the Edinburgh committee's proposals. Both groups manifested their support by means of numerous petitions to Parliament in favour of nationalization.[35]

Those against a state takeover began to give public manifestation of their positions only after the Post Office's declaration of intent in December 1866. So while the backers of nationalization had been building up anti-oligopolistic protest over a ten-year period, the defenders of the status quo began their campaign when most public opinion was against them. In addition, the fact that the press

---

wise and a very large thing, in doing which they will, unless we are greatly mistake, have the cordial support of Mr. Gladstone, the originator of the design. According to the *Observer*, a bill, drafted some time since, is to be introduced this session giving the Postmaster General power to purchase, work and extend all telegraphs in Great Britain and Ireland", "Government and the Telegraph", *The Daily Post*, 26 November 1867.

[33] BPP, 1867–1868, Vol. 41, *Supplementary Report to the Postmaster General upon the Proposal for Transferring to the Post Office the Control and Management of the Electric Telegraphs throughout the United Kingdom*.

[34] *Dundee Advertiser*, 15 August 1865; "A News Telegraph Company", *Glasgow Herald*, 16 August 1865; "The Press and the Telegraph Company", *Leeds Mercury*, 18 August 1865; "The Necessity of Telegraphic Reform", *Dublin Evening Mail*, 18 August 1865; "The Telegraph Wires and the Press", *Sheffield Daily Telegraph*, 19 August 1865; "The Press and the Telegraph Companies", *Nottinghamshire Guardian*, 25 August 1865; "The Evil of Telegraph Monopoly, and the Remedy", *Northern Whig*, 26 August 1865; "The Telegraph Companies", *Pall Mall Gazette*, 3 October 1865; "The Necessity of Improved Telegraphic Communication", *Dublin Evening Mail*, 18 October 1865; "The Tyranny of Telegraphs", *Westmorland Gazette*, 21 October 1865.

[35] According to Post Office documents, 32 chambers of commerce and 297 newspapers petitioned Parliament in favour of nationalization. Perry feels however that the figures are not reliable as many of the protests were solicited by Post Office officials. Perry, *The Victorian Post Office*, pp. 104–105.

had taken sides in favour of nationalization meant that the opposition had found little space in the newspapers, and its voice was only really heard during the parliamentary discussion of the 1868 Telegraph Bill, when it was aiming more at a high indemnification than at getting the Bill rejected.

The opposition fell into two main groups: the railways and of course the telegraph companies themselves.[36] The railways had at least four reasons for opposing nationalization: many of their major shareholders held substantial quotas of telegraph company capital and many of them were directors; almost all of them had drawn up conventions with the principal telegraph companies, which gave them a guaranteed income and the maintenance of their own telegraph wires in exchange for land rights;[37] many of them carried out their own telegraph service for regulating rail traffic; they feared that once the telegraph service had been nationalized, the next victims of Government expansion might well be the railways themselves.[38]

---

[36] As in the case of the backers, the opposition moved into action with petitions against the proposal for nationalization. There were 19 petitions from railway companies, 11 from telegraph companies, and a good 329 from individual shareholders. Perry doubts these figures because he holds that the shareholders did not send individual petitions spontaneously but were solicited to do so by the managers of their companies, Ibid.

This seems to be corroborated for the Electric by the declarations of its directors, who explain it in terms of coordinated shareholder action. According to Grimston, "the proprietors may rest assured that the Directors will defend their property to the utmost and if circumstances should occur and we should be compelled top art with our property we should take an early opportunity of calling the proprietors together and making some recommendation to them". Bidden said, "I defy any Gentleman or anybody else to permit out (?) one single piece of business of a mechanical nature that government has ever conducted satisfactorily or with ever moderate success". Both extracts from BT, *Electric and International Telegraph Company*, TGE 1/2, *Half Yearly Meeting February 1868*.

[37] "At present the Electric Telegraph Company has contracts with almost every Railway Company, by which WAY-LEAVES are afforded by the railways in consideration of works done upon their lines, and other facilities afforded them. In the event of the Government being unable or unwilling to enter into the manufacturing business necessary for works of construction and maintenance, or in the event of their not considering it desirable to carry on the Railway intercourse of the Kingdom now conducted by Telegraph, it is only reasonable to suppose that the railway companies will require *a rental* for the use of the railways over which the lines of Telegraph are passed", *Government and the Telegraphs. Statement of the Case of the Electric & International Telegraph Company against The Government Bill for Acquiring the Telegraphs*, London, Effingham Wilson, 1868, pp. 84–85.

[38] This fear was alimented by declarations from some of the backers of nationalization. "The chief economical and administrative conclusions which I have now to submit

In his 1866 report Scudamore had totally ignored the interests of the railway companies in the telegraph system and in calculating the financial cost of acquiring the network, he had not taken into consideration at all either the lines used exclusively by the railway companies or the rent coming in for allowing the telegraph companies use of their land.[39] Consequently, not only did the 1868 Bill not specify if the railway companies could still use their own lines, but it also did not take into consideration the possibility of compensation for losing rents and free line maintenance.[40] For this reason the railway

---

are: That communication by railway forms part of a connected system, for the transit of persons, letters, information, and goods. That as a regulating, and predominant controlling function, constantly acting, to ensure punctuality in the departure and arrival of trains, as well as the postal deliveries connected with the trains, the postal function has properly a chief place. ... Unity of management, under a public authority of railway communication, will be in itself an important advance in public improvement. Unity of management of telegraphic communication in connection with the post will also be in itself an important and much-needed improvement. But jarring action will be avoided, the improvement in the pecuniary and other results will be the more complete and speedy. If the requisite combination and subordination of administrative functions be considered and provided for at the outset. ... If I may use a simile derived from sanitary science, I would say that, to give entire freedom for the main arteries of communication, to complete the capillaries of the system, the branch railways, to provide for it, as it were, a new set of nerves of quickened sensation and motion, by a cheap system of postal telegraphy, would more than any other measures, put the body politic in a condition of healthy and prosperous action, such as has been imparted to Belgium by the like means", Edwin Chadwick, "On the Railway Reform in Connection with a Cheap Telegraphic Post and a Parcel Post Delivery", *Journal of the Society of Arts*, 15/775, 18 October 1867, pp. 717–728; "A little experience is worth much argument; a few facts are better than any theory; the Government manages the Post Office with success; by a great reduction of charges it has created a vast business, and earns a satisfactory revenue; the Government has purchased and successfully reorganised the telegraphs, and is making them pay; therefore the Government ought to buy the railways, and we should then have railway fares reduced to a third of their present amounts, trains very regular, and accidents few or none. Such are briefly stated, the reflections which have led many persons to join in an agitation, lately increasing, to induce the Government to undertake the gigantic task of acquiring, reorganising, and even working the whole system of railway conveyance in this kingdom. Although many other reasons, of more or less weight, are given for the change advocated, I believe that the main argument, consciously or unconsciously relied upon, is, that *because* the State Post Office and State telegraphs succeed, *therefore* State railways would succeed", Jevons, "The Railways and the State", in Jevons, *Methods of Social Reforms*, p. 353.

[39] BPP, 1867–1868, Vol. 41, *A Report to the Postmaster General upon Certain Proposals which have been made for Transferring to the Post Office the Control and Management of the Electric Telegraphs throughout the United Kingdom*.

[40] BPP, 1867–1868, Vol. 41, *Supplementary Report to the Postmaster General upon the Proposal for Transferring to the Post Office the Control and Management of the Electric Telegraphs throughout the United Kingdom*.

companies emerged as the strongest lobby against nationalization in early 1868. Nevertheless, their opposition was destined to weaken, mainly thanks to intense diplomatic activity by Post Office officials.

The reasons of the telegraph companies emerged in a pamphlet which made an explicit appeal to businessmen, the press and the railways to oppose the Bill. It tried, for example, to play on the fear of businessmen and especially the railways of further nationalization:

> in this country... the people have not been accustomed to subject themselves to Government control, or to rely upon the State to do that for them which they are able to do for themselves. If this principle is a right principle in regard to one class of enterprise, it is also a right principle with regard to another. It should be maintained as a principle; and it is obvious that any infringement on the principle is liable to be made a precedent for further infringements in different directions.[41]

Furthermore, the same pamphlet conjectures on future scenarios in which Government censure would make life difficult for businessmen and the press:

> What is a telegram? Practically it is *an open letter*, the contents of which is known to and is capable of being used by every one through whose hands it passes. It is desirable that the most important part of correspondence of the country should pass through the hands and be subject to the surveillance of Government officials?[42]

As emerges from the above pamphlet, the telegraph companies felt they were the sorely treated victims of abuse and saw nationalization as an act of confiscation.[43] Beyond its purely propaganda value,

---

[41] *Government and the Telegraphs*, p. 68.

[42] Ibid., pp. 78–79.

[43] "After a long and arduous struggle, the Electric and International Telegraph Company has succeeded in establishing telegraphy as a practical science, available, for the most valuable and useful purposes, to all the world. In a single generation, the Company has succeeded in placing Great Britain and Ireland in telegraphic communication with the whole globe, and in placing itself in a position so commercially secure as to render it beyond the danger of competitive attack. At such moment, the Government of the country which they have so greatly served, comes forward to rob them of the fruits of their enterprise, and to take from the founders and originators of telegraphy as a science, the whole benefit of the invention, which, unassisted and unaided, they have carried into practice! Such an arbitrary interference with rights and privileges confirmed by Parliament, is absolutely without precedent in the history of the nation", Ibid., p. 20.

the pamphlet brings out at least three weaker elements in the proposal. First of all, as in its war with the United, the Electric had argued that a low flat charge was not remunerative and that it would not be able to generate the profits foreseen in the nationalization projects. The figures were based on the examples of Belgium and Switzerland, which had been applying an extremely economical flat charge for years. They, however, were in the middle of continental Europe and enjoyed a very intense traffic of international transit. And unlike domestic dispatches, international telegrams were very expensive and produced revenue that went to cover the deficit generated by national traffic. Great Britain was on the outskirts of the European network and would never have been able to make up for the losses on its domestic service with income from international telegrams.[44] True, up to a certain point, for the Post Office was also calculating that the trans-continental cables, which had started operating in 1866, gave Great Britain a transit function in a global rather than European context, and with a Government monopoly Great Britain could send its own delegates to the International Telegraph Union conferences, where high international tariffs guaranteeing considerable revenue were established.[45]

---

[44] "Upon consideration of all the facts, it will be apparent that, in no essential, in there any analogy between the telegraphy of the United Kingdom and that of Belgium and Switzerland independently of other facts, it is to be remembered that our telegrams are internal almost exclusively, whilst their main service is international and transit. The principal messages of Belgium and Switzerland only cross those countries, whilst in the United Kingdom we have no transit messages except a few for the Atlantic Cable. Messages come from foreign parts to England, and are sent from England to foreign parts; they originate here and come here, but they do not cross the United Kingdom. A very large proportion of these Belgian and Swiss telegrams, are telegraphs in transit from England and to England, and we see that, even including these telegrams in transit through the countries, Belgian and Swiss telegrams are in no greater proportion to area and population than those of the United Kingdom. But in one respect these transit and international telegrams, of which there are none in our country, involve a most important difference. Belgium and Switzerland can make up the deficiencies which arise from losses on internal communication by the surplus derived from transit telegrams. In the United Kingdom, if the Post Office lost money on its internal communication by fixing the rates on the same scale as those of Belgium and Switzerland, it would have no transit on international telegraphy from which to obtain a surplus to supply the deficit", Ibid., pp. 44–45.

[45] Only states running their telegraph service directly could join the international Telegraph Union and take part in its periodic conferences. George Arthur Jr Codding, *The International Telecommunication Union. An Experiment in International Cooperation.* Leiden, E.J. Brill, 1952; Leonard Laborie, "En chair et en normes. Les participants aux

Secondly, the Electric recognized that the fusion of the land-lines would lead to the elimination of tracks, offices and personnel operating competitively in the same places, thereby generating a notable lowering of costs. For this reason, it offered to lead a private monopoly, but was discouraged by the 1863 Telegraph Act, which called for the Board of Trade's authorization for any sale/purchase or transfer.[46]

Finally, the pamphlet considered it a paradox to count on the telegraph service being improved by a public administration that had never run it. And in fact Post Office personnel did not have the technical and scientific competences for carrying out the service,[47] nor could they be replaced by the Post Office officials' great organizational capacity, as so many in favour of nationalization had thought.[48]

---

conférences de l'Union internationale des télécommunications, de sa fondation à sa refondation (1865–1947)", *Flux. Cahiers scientifiques internationaux Réseaux et Territoires*, 74, 2008, pp. 92–98.

[46] "Parliament has uniformly dealt with the Telegraphic communication of the United Kingdom as a thing which, in the interests of the nation, would be best promoted by competition. Not only, therefore, has it passed Acts to facilitate and encourage the formation of competing Companies, but it has absolutely enacted laws to prevent the possibility of those Companies uniting and establishing the MONOPOLY which it is now desired to obtain for those Companies. A clause in the United Kingdom Telegraph Act (1862), actually went so far as to provide that it should not be lawful for that company to sell or transfer any portion of their undertaking to any other company or person, without the direct authority of Parliament to do so, and it even prohibited them from leasing any one of their wires without the authority of the Board of Trade. Parliament, therefore, has taken the most jealous precautions against the establishment of a monopoly of the Telegraph Companies. It has, also, most carefully sought to secure to the public the largest amount of benefit arising from the competition of those companies. ... It has done, in fact, everything in its power, under all former Governments, to promote that very competition and to prevent that very monopoly, which, under the present Government, has been suddenly discovered to be essential to the progress of the Telegraphic system", Ibid., pp. 69–70.

[47] "There seem, therefore, to be abundant reasons against committing the Telegraphs to Government control, whilst there are no reasons whatever in support of the proposal. It is difficult to see what it is proposed to be done by the Government for the public interest; but it here is anything that the Bill proposes to effect it can certainly be as well (and probably much better) effected by the Companies. At no period has the Government ever exhibited any special aptitude for telegraphic work. Experience, certainly, does not show that either in the construction or working of a telegraphic system, a Government department has been peculiarly successful. The history of their arrangements with the 'Mediterranean Extension Telegraph' illustrate that they incline to pay £7,200 a year, rather than have their work done for nothing!", Ibid., pp. 86–87.

[48] "Thus much, however, he [Mr. Scudamore] might say, that though he had no claim to all the good words Mr. Chadwick had said in reference to the management of the Post-Office, he thought they could do more to facilitate communication throughout

6.3   The 1868 Telegraph Act

The Bill for Nationalization took two years to pass through Parliament, and involved first Disraeli's Conservative Government and then Gladstone's Liberal one. It began on 1 April 1868, when Ward Hunt, the Chancellor of the Exchequer, presented in the Commons the Telegraph Bill "to enable the Postmaster-General to acquire, maintain and work the Electric Telegraph in the United Kingdom",[49] and it ended in August 1869 with the passing of the Second Telegraph Act.

In the period between the presentation of the Telegraph Bill in April 1868 and the debate in the Commons two months later, the clash between the supporters and opponents reached its apogee. This confrontation, which saw the railway companies still bitterly opposed to nationalization, was transferred into Parliament, where George Leeman, member for York and chairman of the North Eastern Railways,[50] became a leading light of the opposition. He felt in particular that the Post Office would never be able to run the service as efficiently as the private companies and that nationalization would turn out to be an enormous waste of money.[51] His oratory

---

the country than existing telegraph companies were able to do. He thought he could, at the proper time, prove that the Post-Office could, at much less cost than could the present companies, bring telegraph stations nearer to the public, and at the same time save the public considerable expense – that the Post-Office could, at less cost than the present companies, keep the telegraph open five or six hours a day longer than it now was, and thereby afford the public greater opportunities of sending messages than they now enjoyed; and, by reason of the Post-Office being able to work with one staff, with one set of wires, with one set of instruments, and with one central establishment instead of several, they would be able to effect an important reduction in the charges for messages throughout the whole of the United Kingdom", Chadwick, *On the Economy of Telegraphy*, p. 229.

[49] *An Act to Enable Her Majesty's Postmaster General to Acquire, Work, and Maintain Electric Telegraphs* [31 July 1868], 31–32 Victoria, c. cx.

[50] Alf Peacock, "George Leeman and York Politics 1833–1880", in C.H. Feinstein (ed.), *York, 1831–1981: 150 Years of Scientific Endeavour and Social Change*, York, William Sessions Limited, 1981, pp. 234–254.

[51] "Mr. Leeman (York) strenuously opposed the measure, and moved that it be read that day six months. He argued generally that the telegraph would be better worked by private enterprise than by the Post Office; that the press had no grievance to complain of in relation to the present system; that the telegraphic companies had greatly reduced their rates and done everything to meet the wants of the public; and that if this great machinery were transferred to the hands of Government would create corruption and tyranny to which the English public would never submit. In proof of the latter statements, he cited the case of continental governments, who, having got possession of the telegraph wires, had used them to the prejudice of the people", *Sheffield Daily Telegraph*,

did not manage to move the majority of the house, clearly which was on the other side.[52] Nevertheless, his argument insinuated doubts about the financial aspects of the operation, especially among the Liberals sitting on the opposition benches.[53] So, it was decided to

10 June 1868; "The opposition to the Bill was led by Mr. Leeman, who took a fundamental objection to the assumption by the Executive of duties properly belonging to and more efficiently to be discharged by private enterprise, and complained grievously that the Government, by not bringing in their measure as a private Bill, had deprived the companies of the ordinary opportunities of defending themselves. He dwelt forcibly on the inconveniences to the railway companies of putting the telegraph under Government control, asserting that so complicated and numerous were the relations between the two classes of companies that the railways could hardly be worked if this transfer were carried out. Referring to what he described as the 'notorious corruption' of telegraph administration in foreign countries, he drew an alarming picture of the espionage and abuse of confidence which might be the result of giving the Post-office the working of the telegraphs; and he maintained that the Government could not do the work without a considerable loss. He defended warmly the administration of telegraph companies, insisting that they had given the public every possible advantage of economy and efficiency, and he derived the expectation that the system could be greatly extended and particularly the idea that the use of the telegraph would become common in the smallest villages", "The Electric Telegraphs Bill", *Gloucester Journal*, 13 June 1868.

[52] "The government will insist on being victims. They are never honestly opposed. Directly a powerful member of the House of Commons interferes with one of their plans, they condemn his opposition as captious. If there be a question, which is not a party question, in the settlement of which no set of politicians need be compromised – it is surely is the bill which empowers the government to treat with the electric telegraph companies for the purchase of the lines, and authorises the Postmaster general to work the lines so purchased. ... The Wholesale objection of Mr. Leeman defeats itself. We suspect the temper and judgement of the man who will not see the smallest point of good in the long-pondered plan o fan opponent who is his full equal in intelligence. We look in vain for the ground on which Mr. Leeman opines that the transfer of the telegraphic system of the country to government, will leave railway companies unable to work their lines with safety. The deprivation of electric means of Communications, for the government of their ways Could not be accepted by railway companies. The minister who should propose such a scheme, would be forced to accompany it with another for the transfer of railway property to the state; for no board of directors would undertake to regulate traffic without an unfettered telegraphic communication along their line. In seeking obstructions to the government measure in this direction, Mr. Leeman was far astray. The weakness of the measure is elsewhere. The uncompromising opponent should have dwelt on the vices of the French system", "The Electric Telegraphs", *Lloyd's Weekly Newspaper*, 14 June 1868.

[53] "Mr. Leeman said he was willing to withdraw his amendment, but had expected the hon. member for Pontefract would have moved his resolution that a select committee should consider the question of the monopoly of the working of the telegraphs by the Post Office. The Chancellor of the Exchequer said that if the bill were read a second time he would consent to refer it to a select committee, which should receive an instruction to inquire as to whether the transmission of messages should be a legal monopoly in the hands of the Post Office, as to the making of special contracts, as to the secrecy of

appoint a parliamentary committee to go into the bases on which to create the Government monopoly; establish the criteria with which the Post Office could offer telegrams at economical prices; draft regulations for respecting privacy; decide what to do with submarine cables belonging to land companies; listen to those who had petitioned against the Telegraph Bill.[54] What was clear was that Parliament took for granted the approval of the public acquisition of the telegraphs and believed the committee would only have to clarify the details. Designated on 23 June 1868, it included George

---

messages, as to the working of the submarine cables, and as to the hearing of parties by counsel before the committee. Mr. Bouverie did not think this subject had been sufficiently sifted. Mr. Alderman Lawrenoe did not think the public interests ought to be handed over to a department. A parliamentary inquiry ought to have preceded this movement. After a few words from Mr. Phillips, Mr. Norwood said his constituents were in favour of this bill. Three-fourths of the provincial press were in flavor of the bill, but they had been threatened by the telegraph companies in consequence. A Belfast paper had been threatened with the discontinuance of the telegraphic service if it persisted in its support of the bill. He thought the newspaper proprietors should be protested in their receipt of telegraphic communications. Mr. Ayrton desired to preserve the power in the hours of deciding on the question after the select committee had reported. Mr. Newdegate thought it important to guard against the government obtaining a dangerous monopoly in their control of telegraphic communications. Mr. Maguire said that the string of subjects to be inquired into as suggested by the Chancellor of the Exchequer could not be settled for months. He could not understand how the second reading could be had without affirming the principle of the bill. He moved the adjournment of the debate. Mr. B. Carter asked the Chancellor of the Exchequer who would represent the public before the select committee. The Chancellor of the Exchequer said the committee would examine parties in the interest of the public. The public interest would be considered as well as private interests. The amendment was withdrawn, and the bill was read a second time. On the motion of the Chancellor of the Exchequer the bill was referred to a select committee", extract from the parliamentary debate in "Imperial Parliament", *London Standard*, 19 June 1868.

[54] "The Chancellor of the Exchequer moved that it be an instruction to the Committee on the Electric Telegraph Bill to inquire – 1st whether it is desirable that the transmission of messages for the public should become a legal monopoly in the Post Office; 2nd, whether it should be left to the discretion of the Postmaster General to make special agreements for the transmission of messages or news at reduced rates; 3rd what securities should be taken for insuring the secrecy of messages transmitted through the post-office; 4th, what arrangements should be made for the working of submarine cables to foreign countries; and 5th to bear such telegraph and railway companies and proprietors as shall by petition, on or before the 26th instant, pray to be heard by themselves, their counsel, or agents against such of the matters referred to the committee as affect their particular interests. He also proposed that the Select Committee should consist of the Chancellor of the Exchequer, Mr. Goschen, Sir F. Heygate, Mr Leeman, Mr C. Turner, Mr. Norwood, and five members to be added by the Committee of Selection" extract from parliamentary debates, in "By Special wire and ordinary telegraph", *Glasgow Herald*, 24 June 1868.

Leeman and George Goschen, member for the City of London, head of the Früling and Goschen Bank, a director of the Bank of England and former vice-president of the Board of Trade.[55] Leeman, as a representative of the railway interests, and Goschen, for the London businessmen, were there to defend the private telegraph companies from Government interference. Nevertheless, when on 1 July 1868, the Inquiry Committee met for the first time, Scudamore's diplomacy had already done its work, and in little less than a month both railway and telegraph companies had reached advantageous agreements for the cession of their rights and structures. Right from the end of April, the representatives of the Post Office were holding meetings with the Telegraph and Railway managers to negotiate the terms of the acquisition. The telegraph companies were claiming an indemnity equal to their net profits over the previous 20 years, while the Post Office held it was enough to pay the highest prices reached by the shares of all the companies on the London Stock Exchange before 23 May.[56] At the same time Scudamore also declared officially that he was going to offer the railway companies the same conditions they enjoyed in a regime of free trade.[57] Consequently, the committee could not cross-examine any witness against nationalization since the telegraph and railway companies had accepted negotiation with the Post Office and had therefore given up any form of public opposition.[58] In spite of Goschen's and Leeman's intention to

[55] Arthur D. Elliot, *The Life of George Joachim Goschen, First Viscount Goschen, 1831–1907*, London, Longmans Green, 1911; Thomas J. Spinner, *George Joachim Goschen: The Transformation of a Victorian Liberal*, Cambridge, Cambridge University Press, 1973.

[56] Perry, *The Victorian Post Office*, pp. 106–107.

[57] "If the Post Office acquires the property of the Telegraph Companies it will acquire any privileges which they enjoy under their contracts with the Railway Companies, and will be charged with any obligations towards those Companies, Which the contracts may impose. I have made careful inquiry on this subject, and have been unable to discover any reason for thinking that the Post Office may not work at least as well with the Railway Companies as the Telegraph Companies have worked with them", BT, *Electric and International Telegraph Company*, TGE 1/6, *Electric Telegraphs Bill, Return to an Order of the Honourable the House of Commons dated 14 May 1868* [Mr. Scudamore's reply to the pamphlet entitled "The Government and the Telegraphs"], p. 16.

[58] For example, "The Select Committee upon this bill resumed the examination of witnesses on Monday, in N. 11 Committee-room. In the chair was the Chancellor of the Exchequer. A number of counsel were present, but the opposition having been virtually withdrawn they simply watched the proceedings. – Counsel announced that the London and South Western and the Submarine Companies had agreed to terms with the post offices. – Mr. Rodwell said agreements had been actually signed in the majority

make life difficult for Scudamore, in the end, the committee decided unanimously on the advisability of nationalizing the telegraph service, with the only recommendation, shared by the Post Office, that a monopoly not be created.[59] This condition would be respected by leaving the telegraph services the alternative of selling out to the Post Office or carrying on and setting up in direct competition. The Telegraph Bill became law on 31 July 1868 without really arousing any animated discussion during the Commons debate.[60]

Obviously, the new Act appeared to be a compromise between the contenders: on one side it allowed the Post Office to go ahead with the acquisition of the telegraph networks belonging to private companies,[61] and on the other it compensated adequately both

---

of cases with the companies, and in other instances agreed to", "The Electric Telegraphs Bill", *Huddersfield Chronicle*, 18 July 1868.

[59] "The Chancellor of the Exchequer, in moving the committal of the Bill, explained at length what had been done before the Committee. The opposition of the Electric Telegraph companies had been withdrawn on the agreements that they were to receive twenty years' purchase, calculated on the present year's profits. A similar arrangement had been made with the railway companies, who were to be compensated for their reversionary interests at the same rate, and for the cession of permanent way-leaves over the lines. The details of these arrangements – which he maintained, if liberal, were fair – would have to be carried out by arbitration, and, therefore, it would be impossible to bring in the Money Bill until next session. The total sum necessary for the purchase and extensions was now estimated by Mr. Scudamore – and the soundness of his calculations were vouched for by Mr. Foster, of the Treasury – at £6,000,000. ... As a security that the Post-office would do the work well, and would take advantage of all improvements in telegraphy, the Committee had determined to give it no legal monopoly. ... Mr Goschen admitted that the inquiry before the Select Committee had greatly modified same of the original objections to the Bill, and that, on the whole, it had shown the administrative difficulties in the transfer to be less than were anticipated. Mr. Leeman canvassed Mr. Scudamore's figures in a very hostile spirit contending that nearer £10,000,000 than £6,000,000 would be needed to carry out the plan", "Imperial Parliament", *Hampshire Advertiser*, 25 July 1868.

[60] "The Bribery and the Electric Telegraph Bill – two of the most valuable measures passed by Parliament for many years – have been successfully piloted through both Houses and received the Royal Assent", *Royal Cornwall Gazette*, 30 July 1868.

[61] "4. It shall be lawful for Her Majesty's Postmaster General and he is hereby authorized, with the Consent of the Lords Commissioners of Her Majesty's Treasury, from Time to Time, out of any Monies which may be from Time to Time appropriated by Act of Parliament and put at his Disposal for that Purpose, to purchase for the Purposes of this Act, the whole, or such Parts as he shall think fit, of the Undertaking of any Company, and any Undertaking, and all other Property purchased under the Powers of this Act, shall be vested in and held Majesty's Postmaster General, in his corporate Capacity, and his Successors: Provided always, that no such Purchase be made, and that no Agreement other than the Agreements confirmed by this Act for any such Purchase be binding, unless the said Agreement, accompanied by a Minute from the Commissioners of Her Majesty's Treasury, in which the Grounds of the Agreement shall

the lobbies backing nationalization and those initially opposing it. In fact, like a normal contract the text of the law detailed the compensation that the Post Office had offered to those it had had to silence and those who had been in its favour. Article 8, for example, provided meticulously for compensating the Electric, Magnetic and United in the case of selling their networks.[62] Later on, in Article 9, the compensation for the railway companies was given in great detail, including the indemnity both for the loss of the telegraph service where they ran it directly and also where they lost income and

be set forth, shall have lain for One Month on the Table of both Houses of Parliament without Disapproval. 5. Any Company, with the Authority of Two Thirds of the Votes of their Shareholders present in person or by proxy at a General Meeting of the Company specially convened for the Purpose, may sell all or any Portion of their Undertaking to the Postmaster General for such Sum of Money as may be mutually agreed upon between the Postmaster General and the Company; and the Execution by any Company under their Common Seal of a Conveyance to the Postmaster General, duly stamped, of their Undertaking, shall be sufficient to vest the same in the Postmaster General for all the Estate, Right, Title, and Interests of the Company therein, with all incidental Rights, Privileges, and Easements, and the same may be used, exercised, and enjoyed by the Postmaster General in the same Manner and to the same Extent as the same respectively are, or if this Act had not been passed might be held, used, exercised, and enjoyed by any Company, and the Receipt of Two of the Directors of any Company for the Purchase Money endorsed upon the Deed of Conveyance, shall be a sufficient Discharge for the same to the Postmaster General, who shall not be bound to see the Distribution thereof", *An Act to Enable Her Majesty's Postmaster General to Acquire, Work, and Maintain Electric Telegraphs* [31 July 1868], 31–32 Victoria, c. cx, Articles 4 and 5.

[62] "With respect to the Purchase of the Undertakings of the Electric and International Telegraph Company, the *British and Irish* Magnetic Telegraph Company, and the *United Kingdom* Electric Telegraph Company (Limited), enacted as follows: 1. Each of the Three Companies may, with the Authority of Two Thirds of the Votes its Shareholders, present in person or by proxy at a General Meeting of the Company specially convened for the Purpose, sell and convey, and the Postmaster General shall upon demand of the Company under its Common Seal purchase, the whole Undertaking of the said Company. 2. the Price to be paid by the Postmaster General to each Company for its Undertaking shall be Twenty Years Purchase of the net Profits during the Year ending on the Thirtieth day of June One thousand eight hundred and sixty-eight from the Undertaking so conveyed; and in the Case of the United Kingdom Company there shall be paid in addition to the Amount aforesaid – First, the Price paid by the Company for the Patent of 'Hughes' Type-printing Telegraph', such Price not to exceed Twelve thousand pounds; Secondly, a Sum equal to the estimated aggregate Value of the quoted Ordinary Share Capital of the Company reckoned on the highest Quotation shown in the Official Lists of the London Stock Exchange of any Day between the First and Twenty-fifth Days of June One thousand eight hundred and sixty-eight; and Thirdly, compensation for the Loss of the prospective Profits of the Company on the ordinary Shares, and any Sum that may be determined upon consideration of the Efforts made by the Company to establish a uniform Shilling Rate for the Conveyance of Telegraphic Messages", Ibid., Article 8.

various rights coming from conventions established with the telegraph companies.[63] Lastly, Article 16 took count of the press's pronationalization campaigns and granted it privileged conditions for sending its own telegrams.[64] The same article also gave official sanction to a treatment of parity between the various daily newspapers and agencies, since it "provided also, that no such Proprietor, Publisher, or Occupier shall have any undue Priority of Preference

---

[63] "The Postmaster General shall pay the Railway Company the following Sums by way of Compensation:a. Twenty Years Purchase of the Amount of the net annual Receipts (if any) of public Telegraph Messages received and forwarded by the Railway Company on their own Account, reckoned on the Basis of the Receipts derived there from over a continuous Period of Twelve Months prior to the Thirtieth Day of June One thousand eight hundred and sixty-eight; b. Twenty times the Amount of the estimated annual Increase, calculated upon the average Increase of the preceding Three Years of the said Receipts from Telegraphic Messages, or where the Business has been commenced within Three Years calculated upon the Increase during such short Period, such annual Amount in case of Difference to be settled by Arbitration; c. All rents and annual or other Payments payable to the Railway Company by public Telegraph Companies during the still unexpired Periods embraced in their respective Agreements, and at the Terms mentioned in said Agreements respectively; d. Such sums as shall be agreed upon, or in default of Agreement as shall be settled by Arbitration, in respect of the Loss by the Railway Company of the Privilege of granting other Way leaves and making future Arrangements with Telegraph or other Companies, and in respect of granting a Monopoly to the Postmaster General for the Conveyance of Telegraphs over their Railways as herein provided for; e. Such Sums as shall be agreed upon, or in default of Agreement as shall be settled by Arbitration, as the Value of the Railway Company's reversionary Interest (if any) in the Telegraph Receipts from public Messages on the Expiration of the Agreements with the respective Telegraph Companies; f. Such sums as shall be agreed upon, or in default of Agreement as shall be settled by Arbitration, for the Loss occasioned by Removal of any Clerks now provided by the Telegraph Company may incur in working their Telegraph for Railway Purposes as a separate System", Ibid., Article 9.

[64] "Notwithstanding anything in this Act, it shall be lawful for the Postmaster General, with the Consent of the Commissioners of Her Majesty's Treasury, from Time to Time to make Contracts, Agreements, and Arrangements with the Proprietor of Publisher of any public registered Newspaper, or the Proprietor or Occupier of any News Room, Club, or Exchange Room, for the Transmission and Delivery, or the Transmission or Delivery of Telegraphic Communications at Rates not exceeding One shilling for every Hundred Words transmitted between the Hours of Six p.m. and Nine a.m. to a single Address, and a Rates not exceeding One shilling for every Seventy-five Words transmitted between the Hours of Nine a.m. and Six p.m. to a single Address, with an additional Charge of Twopence for every Hundred Words, or Twopence for every Seventy-five Words, as the Case may be, of the same Telegraphic Communication so transmitted to every additional Address: provided always, that the Postmaster General may from Time to Time, with the like Consent, let to any such Proprietor, Publisher, or Occupier the special Use of a Wire (during such Period of Twelve Hours per diem as may be agreed on) for the Purposes of such Newspaper, News Room, Club, or Exchange Room, at a rate not exceeding Five hundred Pounds per Annum", Ibid., Article 16.

in respect of such Rates over any other Proprietor, Publisher, or Occupier".[65] Thus was bypassed one of the reasons behind the press, especially at the local level, supporting nationalization, i.e., alleged disparity in treatment. Not only did the Telegraph Act detail compensation for all those involved,[66] but it also foresaw some of the minimum conditions that the new telegraph service would have to guarantee: a flat charge of 1 shilling for a 20-word telegram; names and addresses not to be included in the 20 words; free messenger delivery for addresses within a mile of the office,[67] and the introduction of telegraph stamps.[68]

In the eyes of the legislators, these prescriptions were to make the telegraph service cheaper, quicker and more efficient. Nevertheless, they immediately catch the attention for two reasons: they are very detailed and they are administrative – not technical – in nature. With such radical telegraph reform, which provided for the state's purchase of the telegraph networks, it would be reasonable to expect in the first place general indications on the organization of the system, the mechanisms for integrating the various private networks. Instead there are some very specific norms closer to in-house regulations than a law with a general character like a Telegraph Act. Secondly, it is also surprising that the norms for a new telegraph service are

---

[65] Ibid.

[66] Articles 10, 11 and 12 of the Telegraph Act, provided for some forms of compensation also for Reuters and the companies managing canals and private roads. Ibid.

[67] "Notwithstanding anything in this Act, it shall be lawful for the Postmaster General, with the Consent of the Commissioners of Her Majesty's Treasury, from Time to Time to make Contracts, Agreements, and Arrangements with the Proprietor of Publisher of any public registered Newspaper, or the Proprietor or Occupier of any News Room, Club, or Exchange Room, for the Transmission and Delivery, or the Transmission or Delivery of Telegraphic Communications at Rates not exceeding One shilling for every Hundred Words transmitted between the Hours of Six p.m. and Nine a.m. to a single Address, and a Rates not exceeding One shilling for every Seventy-five Words transmitted between the Hours of Nine a.m. and Six p.m. to a single Address, with an additional Charge of Twopence for every Hundred Words, or Twopence for every Seventy-five Words, as the Case may be, of the same Telegraphic Communication so transmitted to every additional Address: provided always, that the Postmaster General may from Time to Time, with the like Consent, let to any such Proprietor, Publisher, or Occupier the special Use of a Wire (during such Period of Twelve Hours per diem as may be agreed on) for the Purposes of such Newspaper, News Room, Club, or Exchange Room, at a rate not exceeding Five hundred Pounds per Annum", Ibid., section 15.

[68] "The Payments to the Postmaster General for the Transmission of Telegraphic Messages from one Place to another within the United Kingdom shall (except for Porterage) be made in all Cases by means of Stamps", Ibid., Article 18.

exclusively administrative and contain no indications – not even general ones – of the technology to be adopted.

The fact that in the Telegraph Act there is no reference to technologies signals very clearly that the Post Office officials who were the original drafters of the law did not possess the necessary know-how to run such a technologically complex service as the telegraph. A telegraph service is different from a postal one in that it calls not only for administrative-accountancy skills, but also specialized scientific knowledge, especially physics. Furthermore, including in the Telegraph Act specific norms about tariffs and administrative matters followed the logic yet again of pleasing those who had backed nationalization above all to obtain a reduction in prices, just like the chambers of commerce.

The 1868 Telegraph Act finished with a regulation that risked making the law inapplicable:

> In case no Act shall be passed during this or the next Session of Parliament, putting at the Disposal of the Postmaster General such Monies as shall be requisite for carrying into effect the Objects and Purposes of this Act, the provisions contained in this Act or in the Agreements hereby confirmed relating to the Arrangements with Railway and Telegraph Companies, and all Proceedings there under, shall become void, and the Postmaster General shall thereupon pay to the several Companies mentioned in such Clauses or Agreements all reasonable Costs and Expenses (if any) properly incurred by them respectively in relation to any Proceedings taken under this Act.[69]

Before implementing nationalization, Parliament would have to approve with a new law the overall sum to acquire the private telegraph networks and the rights to exploit them economically. In order to verify precisely what the sum was to be paid out, an inquiry commission was set up, composed of ten members all from the audit section of the Post Office.[70] Its task was daunting:

---

[69] Ibid., Article 24.

[70] The committee members were chosen among the Post Office managers because it was clear they would have to carry out a strenuous task in a short time. Scudamore knew he could impose on his staff rhythms he could not have requested from accountants outside the Post Office. "It is necessary that this task should be performed rapidly and therefore several officers must be employed upon it and must divide amongst them the accounts", BT, *Post Office*, POST 30/202C, *Transfer of Telegraph Companies to the Post*

it had to examine all the accounts furnished by the telegraph companies, compare them and then verify possible incoherencies on the basis of a technical examination on the lines, materials and service performance carried out by telegraph engineers at the service of the committee.[71] After some months of painstaking controls, the committee managed to produce a report in which it quantified the price of nationalization at around £500,000 less than the figure calculated by the companies themselves.[72] So nationalization would cost the state almost seven million sterling, more than triple the sum originally forecast by Scudamore. The rise in the price was mainly due to the time taken by bureaucracy and politicians. The process of drawing up the plan in the first phase was fairly slow. Scudamore presented his proposal almost a year after receiving the mandate. The project was then further slowed down in 1867, because of the difficulties over the change of Government and the Second Reform Bill.[73] When the Bill finally reached Parliament, the Conservatives were in their pre-election hurry to get it passed. The telegraph and railway companies naturally exploited this inconsistent behaviour. Later, during the April 1868 negotiations, they took advantage of the Government's haste to propose the best possible indemnity: the profits for the period July 1867 to July 1868 multiplied by 20 years. It was no coincidence that the Electric had closed the last financial year with a good profit, so it was able to satisfy its shareholders in two ways,

---

Office: Electric and International Telegraph Company Part 1, Letter from Ives Scudamore to Postmaster General, 18 August 1868.

[71] BT, *Post Office*, POST 8/44, *Diary of Mr Richardson, Member of the Post Office Committee of Inquiry.*

[72] BT *Post Office*, POST 8/43, *Reports by the Committee of Inquiry Regarding Private Telegraph Companies' Accounts 1868–1869.*

[73] In the period, some saw in the slowness of the process a reason for the disproportionate increase in the sum to be paid. "If these were merely the fanciful ideas of a young man of twenty-four, which had little to recommend them, and which had little to recommend them, and which eventually came to nothing, there would be small warrant for reproducing them here. Bu all which is planned in 1856 has come to pass – all that is the net revenue; but, then mine was a scheme for spending, not more than a million and a quarter on postal telegraphs, whereas, thanks mainly to hesitation in adopting some such plan, about £ 11,000,000 have been laid out", Baines, *Forty Years*, pp. 302–303.

with a higher dividend and an automatic increase on the sum to be paid by the Post Office.[74]

6.4 The 1869 Telegraph Act and the passage of the telegraphs to the state

After the Government had passed into the hands of Gladstone and a new Postmaster General, Lord Hartington, had been appointed, Parliament took up the telegraph reform again. The Bill presented in 1869 was complementary to the previous one[75] and contained two main innovations: the exact sum to be paid for the nationalization of the telegraphs, and the introduction of a Government monopoly for the supply of the service.[76]

First of all, the State would be confirming the Post Office committee's analysis by paying altogether almost seven million pounds to nationalize the telegraph service; a sum which covered payment to the railway and telegraph companies and the execution of some work on the telegraph lines to uniform the networks into one system. The

---

[74] "Notwithstanding the continued slackness in commercial affairs, the Revenue during the past half-year shows a satisfactory increase as compared with the corresponding period in 1867. ... The Directors have much pleasure in recommending the usual dividend of £ 5 per cent. ... The Government subsequently made an offer to the Company to give them '20 years purchase of the present net profits', of the undertaking, which under the circumstances the Directors accepted, and upon these terms having been agreed to, the Company withdrew from further opposition to the Bill", BT, *Electric Telegraph Company*, TGA 2/1, *Directors' Report*, June 1868.

[75] "This Act may be cited as 'The Telegraph Act, 1869', and this Act and 'The Telegraph Act 1868', may be cited together as 'The Telegraph Acts, 1868–1869'". *An Act to Alter and Amend "The Telegraph Act, 1868"* [9 August 1869], 32–33 Victoria, c. lxxiii, Article 1.

[76] The terms were already contained in the premise to the Act: "And whereas agreements have been entered into by or on behalf of the said Postmaster General for the acquisition of the undertakings of the several telegraph companies enumerated in the schedule to this Act, on payment to such companies respectively of the amounts set forth in such schedule, amounting in the whole to the sum of five millions seven hundred and fifteen thousands and forty-eight pounds eight shillings and eleven pence: And whereas in further pursuance of the said Telegraph Act, 1868, the Postmaster General has entered into arrangements with divers railway companies, and negotiations are now pending between him and other railway companies as to their interest in telegraphic business, and it is estimated that the amount which will be required for the purchase by the Postmaster General on behalf of Her Majesty of the interest of such railway companies in telegraphic business will not exceed the sum of seven hundred thousand pounds: And whereas it is estimated that the amount which will be required for the other purposes of the said recited Act and of this Act will not exceed the sum of the three hundred thousand pounds", Ibid., Preamble.

price seemed very high already in the period,[77] but it was perceived as just compensation to the telegraph companies for being deprived of the freedom to carry on with their activity. Scudamore himself was aware that the State was paying more than the real value of the entire system. In fact, when he was called to depose before the parliamentary committee, he admitted that if the Post Office had constructed its own lines, it would have cost far less than nationalization.[78]

In the second place, contrary to the 1868 Select Committee's suggestions and also Scudamore's earlier declaration,[79] the clause pertaining to the Government monopoly was introduced.[80] The reasons for Scudamore's radical change of mind are well summed up in his words before the committee:

---

[77] "The most important bill now on the paper is the Telegraph Bill, a bill to hand over all the telegraphs of the country to the Government, the biggest job that has been done for many years. It would cost, to put up these telegraph wires and to provide the necessary plants, less than two millions. We are to give to the different telegraph companies, for these wires and plants and goodwill of their respective trades, five millions, and also provide two millions more for improvements and extensions. It may be a very valuable whistle that we are buying, but surely we are paying a large price for it. In proof thereof, note the rise of the various telegraph companies shares", "My London Letter", *Liverpool Daily Post*, 6 August 1869.

[78] "311.What is about the sum which will have to be paid by the Government of the purchase of the telegraphs? – about six millions and three quarters; rather less, probably. 312. That sum includes, does it not, something in the nature of goodwill? – Yes. 313, You cannot state exactly how much, probably? – I should think as nearly as possible two-thirds would be in the nature of goodwill; perhaps I am overstating that; it may be something rather less....314. *Chairman*, At all events, a considerable proportion is in the nature of payment of goodwill? – Certainly. 315. Would it be possible to establish an entirely new system of telegraphic communication over the whole kingdom, for a less sum than that six millions and three quarters? – Certainly. 316. Very considerably less? – Certainly; we could establish the whole system for at least what it has cost the companies to establish it", BT, *Electric and International Telegraph Company*, TGE 1/6, *Telegraph Bill Opposition, Report of Select Committee on Telegraph Bill 1869*, questions 314–316.

[79] "296. [Mr. Hunt] The Bill of the last year, proposed no monopoly in the hands of the Postmaster General? – It did not. 297. You were examined last year on that subject? – Yes. 298. And you stated at Question 126, 'So long as we do the work well, with such an organisation as we have, we can defy competition'. Are you of that opinion still, or have you modified it? – I [Mr. Scudamore] still think the Post Office can keep its own against competition, but I have seen reason to think, if we have not the monopoly, we might be subject from time to time to very considerable annoyance, annoyance which would not in the long run prevail over us, but which might for a time subject us to some slight loss", Ibid., questions 297–298.

[80] "The Postmaster General, by himself or by his deputies, and his and their respective servants and agents, shall, from and after the passing of this Act, have the exclusive privilege of transmitting telegrams within the United Kingdom of Great Britain and

At question 2464, Mr. Goschen asked you [Mr. Scudamore], "With regard to the question of legal monopoly, have you given your attention to that?" You say "Yes; and my opinion still is, that the Post Office does not want a legal monopoly. They need not fear competition even in lucrative districts". Are you still of that opinion? – Since the passing of the Bill of last year, I have heard of one or two schemes that were set on foot to pass over most lucrative districts; I do not say those schemes would have succeeded, but probably they would have given us annoyance for some little time; I have heard, for instance, of a scheme to connect the Stock Exchange of London, Liverpool, Manchester, and one or two other towns in the North by wires, for the transmission of Stock Exchange telegrams only. Supposing such a scheme to have been tried, it might certainly have drawn away for a time, a portion of our most lucrative business, but I still believe that in the long run, our superior mode of doing business, and the greater sense of security which the Stock Exchanges would have had in Government management, would enabled us to beat the promoters of the other scheme; I believe, however, that the scheme would have been an annoyance.[81]

The fear that public opinion could back a new privatization, just as it had backed nationalization, had driven Scudamore not only to call for a Government monopoly but also to reject an MP's proposal to grant a seven-year renewal.[82] Once again, in spite of criticism lodged

---

Ireland, except as herein-after provided; and shall also within that kingdom have the exclusive privilege of performing all the incidental services of receiving, collecting, or delivering telegrams, except as herein-after provided", Ibid., Article 4.

[81] BT, *Electric and International Telegraph Company*, TGE 1/6, *Report of Select Committee on Telegraph Bill 1869*, question n° 299.

[82] "303. My proposal [Mr. Hunt] is that the Bill should provide for a monopoly in the Post Office for a limited period, say seven years, so that the Post Office would have the opportunity of coming to Parliament and asking for an extension of that term during the term, by which means you would be protected during the term, from being asked to buy off companies that might be got up, and at the same time there would be an inducement to you to give the public every accommodation in regard to telegraphic communication; have you considered that? – We should be undoubtedly protected through the term, but we ought not only to be protected against any person claiming compensation when we came for a renewal of the monopoly at the end of that term; that ought to be secured, undoubtedly. 304. Supposing is were provided, first of all, that the monopoly should be for the period of seven years and it was provided that, in the event of any Bill being introduced into Parliament during such term providing for the extension of the period of enjoyment of, or for the perpetuation of, such exclusive privileges, no person or company should have any *locus standi* to be heard against such extension or perpetuation; would not that be a complete protection to you? – It would be a complete protection to us against any speculator, but I still doubt whether the onus ought to be thrown on the Government of getting a renewal of its monopoly; I am inclined to think the onus ought to be

by some MPs, the Telegraph Act met little resistance during the committee sittings and even less in Parliament.[83] Thus, on 9 August 1869 the second Telegraph Bill officially became law, thus giving a kickstart to nationalization and a Government monopoly.

Kieve holds that the nationalization of the telegraphs took place outside the Houses of Parliament. He argues that the Post Office officials and the delegates from the telegraph and railway companies defined the particulars of the process, bypassing Parliament and placing it twice in front of a fait accompli.[84] However, though there was little participation in the parliamentary debate, that does not mean that Parliament did not play an important role. Perry, in fact, sustains that the two parliamentary passages with their select committees and publicity in the press witness its active role. The debates on the Telegraph Acts was brief and lacked intensity simply because most members of Parliament were as one with the general public, that the only solution to the problems of the British telegrams was nationalization.[85]

---

thrown on the public, of taking away the monopoly from the Government, which they could do hereafter, even if Parliament were now to give a perpetual monopoly", Ibid., questions 303–304.

[83] The criticism coming from Torrens and Fowler was about the total cost of nationalization: "Mr. Fowler complained strongly of the terms of agreement made with the Telegraph Companies, which, quoting the speeches of Mr. Goschen and other members of the Cabinet last year, he denounced as excessive. He especially complained of the method in which the agreements were entered into, and, although approving of the transfer of the telegraphic Business to the Government, he would rather sacrifice all the advantages than consent to such terms as the present, which involved a waste of two millions of public Money. Mr. R. Torrens moved that the bill be referred back to the select committee, and hoped that it would have the effect of defeating the bill, which, in addition to a present loss of three or four millions, would involve a very heavy annual outlay. He disputed the accuracy of Mr. Scudamore's estimate, and contended that the price ought not to have been more than 2 ¼ millions. Mr. W. H. Smith, as a director of one of the companies, pointed out that they were very reluctantly parting with the property. He showed the vast increasing ratio of the profits, which had risen from £ 62,000 in 1862 to £172,000 in 1868, and the 20 years purchase calculated on that year was only 16 years' purchase on the profits of 1869. He also pointed out that the shareholders had borne all the risk, and a very serious risk it was in the early stages of the undertaking – and as the Government had never desired to purchase it until it was a success, they were entitled to consideration for their increasing profits. He showed also that the Post Office ought to manage the telegraphs with greater economy than independent and rival companies", "Imperial Parliament", *Yorkshire Post and Leeds Intelligencer*, 28 July 1869.

[84] "The direct negotiation between the Post Office and the companies, by-passing Parliament, and, in fact, presenting it with a virtual fait accompli, was damaging to the control it had over the spending of public money", Kieve, *The Electric Telegraph*, p. 174.

The purchase had been affected on January 29, 1870, but the transfer was postponed until February 5. There was no formality, no last attendance of the former proprietors, no breaking of white wands, no yielding up even of the keys of the offices. On the night of the 4th the managing directors of the companies walked out; on the morning of the 5th the officials of the Post Office walked in.[86]

So the handover from the private companies to the Post Office took place very quickly, and the government telegraph system started to function at once, partly because the privates had opposed a soft transition because of the additional expense.[87] As the Post Office plan involved the merger of outlying offices and the elimination of redundant lines, the lightning changeover also created problems in running the service. Many employees found themselves suddenly using different wires, different telegraph circuits and different

---

[85] "It is also a mistake to assume that Parliament was not involved in the decision-making process throughout the passage of the bill. Kieve has asserted that 'the direct negotiation between the Post Office and the companies, by-passing Parliament, and in fact, presenting it with a virtual fait accompli, was damaging to the control it had over the spending of public money'. This judgement unduly minimises the consideration of the proposal by MPs in two separate sessions under two different governments. In a early round of debate Ward Hunt had pointed out the essence of the situation facing Parliament. The two sides involved had simply come to an agreement and brought the agreement before Parliament for sanction or rejection. It could choose either, and Parliament had many opportunities to reject the terms... Rather the easy passage with the large favourable majorities on the amendments indicated strong agreement among private lobbyist, civil servant, and MPs as to the necessity of nationalisation", Perry, *The Victorian Post Office*, pp. 118–119.

[86] Baines, *Forty Years at the Post Office*, Vol. 2, p. 24.

[87] "I have considered the question of more Companies than our transferring their wires apparatus and business simultaneously to the same Post Office, and conducting their business from it. The interests of the Companies being antagonistic there would be obstacles that appear to be insuperable to this plan. The customers sending telegrams are using perhaps one Company for foreign, and another for inland telegrams and sometimes sending either way indifferently. Then there are the chance customers; all there must come to one and the same counter to give in their messages, and it is difficult to all who could decide by which Company's wires the messages should be sent. In periods of pressure even now the messages can barely be counted, coded, checked, charged and sent up to the Instrument room so as to keep the work clear, and the difficulty will be enhanced when all the work has to be done at one counter. Each Company could not be represented by a counter Clerk – each clerk endeavouring to secure business for the Company he represented – and the space allowed indeed would probably prevent this if it were not other inexpedient. ... Resuming the 3 Companies to concentrate in the Post Office, it is difficult to see how 3 counter clerks, 3 clerks in charge, 3 sets of clerks, 3 sending out departments, 3 sets of messengers can be depended upon to work harmoniously together when they are all competing together

regulations for sending telegrams. There followed in fact weeks of great agitation, in which top Post Office officials were forced to put into practice themselves their carefully drawn-up projects. Such a rapid change in system would not have been possible if detailed plans on the functioning of the new system had not been circulating for many months. In February 1869 Baines had given Scudamore a memorandum detailing the reorganization of the telegraph service all over the country. It did not neglect technical details, and even less administrative ones. It provided, for example, for a reorganization of the national network with new central and peripheral offices created out of merging outlying offices belonging to rival companies and partially adapting post offices. It also illustrated which telegraph apparatus to use in the various offices to replace previous equipment.[88] What Baines did was to create a plan for the creation of a single telegraph network by means of a powerful technical and administrative standardization.

To bear witness to the fact that the Post Office officials had scarce specific technical competence in running a telegraph service, Baines's plan was subjected to the careful analysis of an outsider, Robert Culley, chief engineer of Electric at the time. After examining the project in detail, he produced a report containing suggestions

---

in the same building for the business of the public! The Secretaries of the 3 Companies have discussed this question, and subject to the approval of the Boards of the respective Companies, were of opinion that the only practicable working arrangements would be the following: for all the Companies to agree to transfer wires, apparatus and business one after another, as prepared, or all at once when ready, into the Post Office. Each Company to make up a total at all the towns where it competes with either or both of the other companies of its gross traffic up to the hire of the first transfer being made into a Post Office working simultaneously from there with a competing Company. Such total to be calculated from the beginnings of the half year down to the period of the first transfer above stated. An average of this to be worked out at per week for each Station. In consideration of the foregoing arrangements to the transfer simultaneously. Her Majesty's Postmaster general to guarantee to each Company that in the event of its traffic from any competing station for the period from the first transfer to the date of the general transfer of the Companies undertakings to the Post Office, being less than the average above worked out, any sum required to make up the amount to such average shall be paid to the Company by her Majesty's Post Master General", BT, *United Kingdom Electric Telegraph Company*, TGH 2/1/1, Letter from Andrews (secretary of the United) to Scudamore, 9 December 1869.

[88] BT, *Post Office*, POST 30/191C, *Adaption of the Wires of Telegraph Companies to the Post Office System*, F.E. Baines, *Post Office Telegraphs Works. Principles of Redistribution. Memorandum*, 22 February 1869.

and modifications in March 1869,[89] and within a few months was appointed chief engineer to the Post Office. Some months after the material transfer of the telegraph system into the hands of the Post Office, the service started at full speed.

> In the twelvemonth from October, 1869, to October, 1870, the Transfer, there were bought and fixed in position 3,382 tons – equal to about fifteen thousand miles – of iron-wire, nearly two thousands miles of gutta-percha covered copper wire, about one hundred thousand poles and a million of other fittings. Moreover, 3,500 telegraph instruments were obtained, and 15,000 batteries to work them with. Joint stock companies were valued and paid out; leased buildings were taken over and occupied. The Post Office engaged and trained about 2,400 new telegraphists and temporary assistants, and appointed more than 1,600 additional telegraph messengers.[90]

Although Scudamore later extended the telegraph network to many more previously cut-off towns and introduced a flat one-shilling charge, ordinary costs grew well beyond what had been foreseen. A first consequence was the resignation of Scudamore; who was also being criticized for an unauthorized use of public funds to increase the network. A direct consequence was that the cost of the personnel (partly inherited from the telegraph companies and partly transferred from the Post Office) spiralled out of control. It is well known today that one of the main defects of a public monopoly is the ever-increasing rise in the cost of work, often encouraged by the string-pulling strategies of the strongest unions within the structure. If it is true the nationalization of the telegraph service was the first in modern times and therefore there were no previous examples to look to, it is also true that the country had at hand the example of the Post Office itself, which in spite of its efficiency, had been registering an increase in personnel costs for decades. It is surprising, therefore, that during the long and intense debates preceding nationalization, the possibility had never been mooted of rising work costs leading to problems of efficiency.[91]

---

[89] BT, *Post Office*, POST 30/191C, *Adaption of the Wires of Telegraph Companies to the Post Office System, Post Office Telegraphs. Mr Culley's Report*, 15 March 1869.

[90] Baines, *Forty Years at the Post Office*, Vol. 2, pp. 34–35.

[91] Hugo Richard Meyer, *The British State Telegraphs. A Study of the Problem of a Large Body of Civil Servants in a Democracy*, New York, The Macmillan Company, 1907.

6.5   The motives leading to nationalization

The official reasons were partly linked to the definition of a natural monopoly, according to which a monopolistic management of some services allowed them to be extended also to less profitable areas and dispensed with lower prices.[92] The Post Office would manage to achieve these two aims by exploiting the advantages of a natural monopoly with functions as noted by John Stuart Mill,[93] who had had in mind public services like water and gas, which call for the construction of big infrastructures and consequently high fixed costs. As natural monopolies, they allow the economies of scale to be exploited and avoid the duplication of infrastructures on the same territory.[94] In the case of the telegraph, fixed costs were not high, but a monopoly would bring about the elimination of dual telegraph lines set up by rival telegraph companies. Besides, a public monopoly had an extra in that the service would be run by a public administration, and therefore there would be no profits to distribute as dividends. All that was required was to balance the books. In a situation like this, the backers felt that in comparison to private companies, the Post Office would find it easier to lower tariffs and extend the service to less profitable areas of the territory.

In reality, there was a different concept of the telegraph service behind the two official motivations. The private companies had always seen their service as exclusive, in clear contrast with the universal and cheap Penny Post. The concept of a universal

---

[92] The official reasons were those propagated by the backers of nationalization, the chambers of commerce and the Post Office, and included in the preamble to the 1868 Telegraph Act: "Whereas the Means of Communication by Electric Telegraphs within the United Kingdom of Great Britain and Ireland are insufficient, and many important Districts are without any such Means of Communication: And whereas it would be attended with great Advantage to the State, as well as to Merchants and Traders, and to the Public generally, if cheaper, more widely extended, and more expeditious System of Telegraphy were established in the United Kingdom of Great Britain and Ireland, and to that end it is expedient that Her Majesty's Postmaster General be empowered to work Telegraphs in connexion with the Administration of the Post Office", *An Act to Enable Her Majesty's Postmaster General to Acquire, Work, and Maintain Electric Telegraphs* [31 July 1868], 31–32 Victoria, c. cx, Preamble.

[93] Paul Johnson, *Making the Market. Victorian Origins of Corporate Capitalism*, Cambridge, Cambridge University Press, 2010, pp. 25–27.

[94] John Stuart, Mill, *Principles of Political Economy*, Vol. II, Boston, Charles C. Little & James Brown, 1848, pp. 537–557.

service[95] had emerged with the entry of the British in the early 1850s, when it challenged the Electric by promoting a cheap, widespread service it was never able to furnish. Almost ten years later, the idea of universalization came to the fore again, when the United and others[96] offered a low-cost service with a flat charge of 1 shilling, while the Electric and the Magnetic stayed with their idea of a mainly exclusive service.[97]

This opposition between universal and exclusive service was the central point in the debate over nationalization. While Scudamore and the Post Office thought, "The telegraph is after all but a quicker post; the telegram is nothing but a condensed letter, and, other things being equal, the nation which most freely uses the post will also most freely use the telegraph",[98] the managers of the Electric saw the things differently:

---

[95] On the concept of universal service applied to telecommunications and in particular telegraphy see Herbert S. Dordick, "The origins of universal service. History as a determinant of telecommunications policy", *Telecommunications Policy*, June 1990, pp. 223–231.

[96] "The United Kingdom Electric Telegraph Company Limited has been formed to carry out a cheap system of Telegraphic Communication throughout the United Kingdom, at a moderate and uniform rate of charge – one shilling, or an equivalent amount. ... By the adoption of a universal rate of charge, on the admirable principle of the penny post, not only existing telegraphic business be greatly augmented, but a new class of customers will be called into existence. Tradesmen requiring the immediate supply of goods will find it advantageous to spend one shilling, where the margin of profit on orders renders the present rates prohibitive. The simplicity and uniformity of charges will greatly induce persons to telegraph, who are at present deterred by the uncertainty of the cost, as well as by its amount. It is worthy of notice that the reason alleged for the maintenance of the existing excessive charges is that the wires are even now fully occupied with messages, and that a reduction could not therefore be advantageously made. The object of this Company will be not to retain high charges and limit communication, but to lower Telegraph rates, and make provision for the largely increased traffic inevitably resulting from such a measure. Its object is to popularise and bring the Telegraph within the reach of the great mass of the public. This it feels satisfied of accomplishing", BT, *United Kingdom Electric Telegraph Company*, TGH 2/1/1, *Prospectus 1860*.

[97] "We do not pretend to compete with the Post Office that is we are no competitors with the Post Office. The Telegraph is only an adjunct to the Post Office, and does certain work the Post Office cannot perform – and the great object of their system is that every message shall be delivered separately and immediately – If you post your letter it may remain for any length of time – from 2 to 12 hours, and may then be despatched with hundreds of others in a mail train, and the same man who delivers yours will have deliver hundreds of others, but the Telegraph message is sent singly and separately along the wires. ... Our best customers [are] the great merchants and brokers of London", BT, TGE 1/2, *Electric and International Telegraph Company, Half Yearly Meeting February 1861*.

[98] BT, *Electric and International Telegraph Company*, TGE 1/6, *Report of Select Committee on Telegraph Bill 1869*.

The fact is, that Postal correspondence and Telegraphic correspondence are two very distinct things. They are NOT "in the same category", or in anything approaching the same category. A broker in London wants to supply his correspondents in the country with prices and information of various descriptions, on which the success of their daily operations is dependent. He resorts to the instantaneous communication afforded by the Telegraphic wire. On the other hand, a lady in London desires to afford to her country cousins and acquaintances the fullest information upon household, domestic, and family affairs of interest, the particulars of which require to be conveyed in full, but which are not of pressing or immediate importance. She resorts to the cheaper, but less expeditious, machinery of the Post. The Telegraph and the Post have thus, their respective duties to perform: each, no doubt equally important to the public, and each, in their respective ways, of public interest. But the two modes of communication are not "in the same category". The two things are perfectly distinct and separate.[99]

Furthermore,

The fact is that the class of individuals to whom the Telegraph is useful are not to be found amongst the class sought to be supplied by the extension of Electric wires to Village Post Offices. The class who use the Electric Telegraph most freely are stockbrokers, mining agents, ship-brokers, colonial brokers, racing and betting men, fishmongers, fruit merchants, and other engaged in business of a speculative character or who deal in articles of a perishable nature. Even general merchants use the Telegraph comparatively little compared those engaged in the more speculative branches of commerce.[100]

Ironically, the Post continued to determine the nature of the telegraph service. Initially, it had conditioned its form of specialization, and now it furthered its universalization so that it could move it into the Post Office. Telegraph clients and the Post Office had in common a belief in a universalized telegraphy, and thus it became the fundamental principle underlying the nationalization project.

According to Perry, the nationalization of the telegraphs was only one of the moves in the progressive expansion of the Government in the late nineteenth century.[101] This may well be true, but a second motivation derives from international politics, and the constitution

---

[99] *Government and the Telegraphs*, p. 53.
[100] Ibid, p. 62.
[101] Perry, *The Victorian Post Office*.

of the 1865 International Telegraph Union (TU),[102] the first supranational government organization in modern times. As the ancestor
of the present International Telecommunications Union (ITU),
it had the triple scope of guaranteeing tariff uniformity, homoge-
neous norms and technological standardization over the whole
international network.[103] In this way telegraphic communications
between member states would be more efficient. The TU was based
on an extremely advanced organizational model based on a permanent body, very similar to the present general secretariats, as well
as periodic plenary conferences. The conventions and regulations
established during the conferences carried the value of international
treaties, and for this reason only state delegations could take part
with voting rights. It followed that only the nations where the telegraph service was directly run by the government could be members.
Beside its own functions, the TU had a great diplomatic importance
because at the time it was the only example of a supranational body
in which the states gave up their own sovereignty to reach a common
aim.[104] Could the United Kingdom, the top economic and political
heavyweight of the period, afford to stay out of a diplomatic experience of such dimensions?

Though there are no explicit references to the Telegraphic Union in
the official documents, some elements allow us to surmise that the
backers of nationalization were well aware of the question. Initially,
back in the 1850s Great Britain had been left on the margins of the
first international telegraph treaties.[105] Parliament had this well in

---

[102] According to Peter Cowhey, international organizations like the Telegraph Union
were desired by the ruling classes to reinforce the status quo. He argues that the bureaucratic state monopoly of telecommunications (in the case of the Post Office telegraphy)
in alliance with the entrepreneurs exclusively furnishing material for building infrastructures upheld an international system of rules guaranteeing and protecting the
monopoly itself within the various nations. If this theory is applied to the nationalization of the telegraphs, it could well be surmised that the ruling classes and the Post
Office bureaucrats were very much in favour of joining the Union. Peter F. Cowhey,
"The International Telecommunications Regime: The Political Roots of Regimes for
High Technology", *International Organization*, 44/2, Spring, 1990, pp. 169–199.
[103] Codding, *The International Telecommunication Union*.
[104] Patrice Durand Barthez, *Union Internationale des Télécommunications*, *Thèse pour le
doctorat en droit*, *Université de Paris I – Pantheon – Sorbonne Sciences Economiques – Sciences
Humaines – Sciences Juridiques*, 1979.
[105] Simone Fari, Gabriele Balbi, Giuseppe Richeri, "European Multilateralism" (1848–
1865): A Telegraphic Idea? Paper presented at the 6th Plenary Conference of Tensions of
Europe, Paris, September 19–21, 2013.

mind, and in 1853 had already put aside £30,000 to build a state telegraph line between London and the continent, as the only way to adhere to a telegraph convention together with the other European powers and guarantee special rates and conditions.[106] In spite of the allocation of the funds, Great Britain did not sign the treaty. More than ten years later, when the Post Office officially presented the nationalization plan to the treasury, a newspaper underscored at once that thanks to this operation Great Britain could soon join the group of powers that had established the TU.[107]

---

[106] "The next vote was for £ 30,000, for the purpose of laying down an independent line of wires from London to the continent, and thus to enable Her Majesty's Government to join a convention recently concluded between France, Belgium, and Prussia for regulating the transmission of messages by electric telegraph. In reply to a question from Mr. Hume, Mr Cardwell said that a convention had been agreed upon between certain foreign countries for the establishment of electric communication throughout the whole continent of Europe. This country was invited to join the convention, but the government of the day did not accept the invitation because they had no power of establishing a communication with the wires upon the continent. That communication had, however, since been effected, and he thought it most important that the Government of this country should avail itself of the advantages of the convention. The advantages of joining the convention, so far as the diplomatic service was concerned, were the right to priority of intelligence, to send messages, which might be sent through several countries without any delay. Then the advantage with regard to commercial messages was this – the English language was not allowed to be used in the transmission of English messages across the continent, unless the British Government were parties to the convention, messages in the English language might, however, be sent to all the continental countries the Government of which had agreed to the convention. The estimate now proposed included the sum required for laying down six independent lines of wire from London to Dover and thence to the opposite coasts. This was the estimate of the whole cost Parliament would be called upon to incur. He had been informed by the Foreign Secretary that, even if this arrangement were confined to the transmission of diplomatic messages, it would be attended with great advantage. He (Mr. Cardwell) had, however, to deal with the commercial bearing of the question, and he was most desirous that if any just and fair arrangement could be made, the benefits of this mode of communication should be available for the interests of British commerce. He might observe that it was not intended that any of the lines of this telegraph should extend beyond London, or interfere with the perfectly free competition which now existed between the electric telegraph companies whose lines were established in this country. The States which had already joined the convention were France, Belgium, the Netherlands, Hanover, Wurtemberg, the German States, Prussia, Austria, Bavaria, Saxony, Switzerland, and Sardinia. ... The vote was agreed to", "Imperial Parliament", *The Cork Examiner*, 8 August 1853.

[107] "It may be anticipated that, with such a change in the telegraphic system, more liberal terms than any possible in the arrangements of joint stock companies can be entered into with foreign Governments. In time a clearing system may be introduced to facilitate correspondence with the Continent. A new value is given to the scheme by the telegraphic connection that has been effected with the United States, and it is

Furthermore, during a Society of Arts congress, Chadwick, a most articulate supporters of nationalization, highlighted that one of the merits of the operation was the possibility of facilitating and improving communications with the Continent.[108] Finally, Scudamore held up Belgium and Switzerland as the examples to follow, though he made no mention of the TU. He had visited their administrations and was obviously well aware that the two countries were the leaders of the telegraphic multilateralism which had led to the union.[109]

As if to publicize the importance of its entry into the TU, Great Britain started pulling punches right from its first conference in 1871. It insisted on a double vote (one for the home country and the other for India) and exploited the presence of the private submarine companies, all British-financed. It also tried to oppose the proposals of the Belgian and Swiss delegations, in a challenge against their leadership.[110]

A second international motivation was the desire of the British Government to create a "red line" in order to control all the territory in the Empire. In the period in question, Great Britain was in fact the only nation to possess the technology and financial resources to construct long intercontinental telegraph cables. Given that it was aiming to set up a submarine telegraph network connecting all the territories of Empire with the home country without passing

highly probable that, at no distant period, the Government of that country may in like manner, unite the telegraphic and postal systems", "The Telegraphs and the Post Office", *Morning Post*, 10 December 1866.

[108] "About three years ago there was an international congress held in Paris of executive officers engaged in the postal departments of the chief continental states, to consider and promote improvement in international communication by letter. At that congress, England had its place. Last year, a like congress was held of the executive officers of the continental states to impart to each the benefit of the domestic experience of all scientific, mechanical instruments, alphabets, as well as in administration and to promote improvement in intercommunication by the telegraph. A that congress, and on that subject, in which England had the lead in the wonderful application of electricity – one great glory of our time – the English Government, to its European scandal for its torpidity, had no place, and England no representative. Is that to continue to be so?", Chadwick, *On the Economy*, p. 226.

[109] Gabriele Balbi, Simone Fari, Spartaco Calvo, Giuseppe Richeri, "Swiss Specialties: Switzerland's Role in the Genesis of the Telegraph Union, 1855–1875", *Journal of European Integration History*, 19(2), 2013, pp. 207–225.

[110] Simone Fari, *The Formative Years of the Telegraph Union*, Cambridge, Cambridge Scholar Publishing, 2015.

over foreign soil, it was felt such a network had to be run by British companies alone.[111] In this way, the Government wanted to guarantee for itself red lines of communication with its own territories which could not be easily hacked in time of war.[112] The project was terminated in the 1890s, thanks to the considerable contribution of John Pender.[113]

Following the exemplary cases of Pender and Thomas Brassey, some scholars have conjectured that one of the main reasons underlying the nationalization of British landline telegraphs was to free capital to invest in submarine companies.[114] In this way Government and Parliament are thought to have indirectly favoured the construction of the desired services wholly in British hands. This would justify the need to free financial resources to favour the expansion and completion of the red lines.[115] Furthermore, the hypothesis of an indirect incentive, rather than direct help in building links with the dominions, seems to be confirmed by the failure of the subsidies policy for submarine companies adopted in the famous cases in the 1860s.

A possible weakness lies in the difference between landline and submarine telegraph companies. Between 1850 and 1860, the former were a fairly solid medium-to long-term investment: in fact, main landline companies like the Electric and the Magnetic yielded quite high profits for almost a decade while the submarine companies were notoriously high-risk investments, as the first transatlantic and Red Sea cables had shown.

---

[111] Daniel Headrick, *The Invisible Weapon. Telecommunications and International Politics 1851–1945*, New York, Oxford University Press, 1991, pp. 20–24.

[112] Robert Boyce, "Submarine Cables as a Factor in Britain's Ascendancy", in Michael North (ed.), *Kommunikationsrevolutionen. Die neunen Medien des 16. und 19. Jahrhunderts*, Köln-Weimar-Wein, Böhlau Verlag, 1995, pp. 81–100.

[113] Peter J. Hugill, *Global Communications since 1844. Geopolitics and Technology*, Baltimore, Johns Hopkins University Press, 1999, pp. 25–52.

[114] "Something like £8,000,000 sterling for re-investment by those who looked favourably on electric telegraphs as a subject of safe and sure remuneration", Bright, *Submarine Cables*; "When the British government bought out its domestic telegraph companies for £8 million in 1869, former shareholders were eager to invest their newly liquid capital in other telegraph enterprises", Headrick, *The Invisible Weapon*, p. 33; "Perhaps the biggest benefit to cable investment was the sudden release of British capital made possible the British government's nationalization of the domestic telegraph companies in 1868", Hugill, *Global Communication*, p. 32.

[115] Paul M. Kennedy, "Imperial Cable Communications and Strategy, 1870–1914", *The English Historical Review*, 86/104, October, 1971, pp. 728–752.

Nevertheless, something had changed in the early 1860s. The technology of building, laying and working underground cables had radically improved and breakages were almost nonexistent. As Hempstead brilliantly describes, the engineers' and technicians' painstaking and cutting-edge performances emerge very clearly from their very detailed reports of their work in progress, which awoke public/shareholder interests via their conferences.[116] After the dual success of the submarine cables in 1866, investors had understood the submarine company shares were no longer high risk but went on chalking up notable levels of returns. As we saw in the last chapter, submarine cables went through a moment of great development while landline technology stagnated. The situation worsened after 1865, when the process of nationalization began to gather momentum. The prospect of the telegraph network passing into the hands of the state made the telegraph companies draw back from investing in infrastructures and engineers, and technicians abandoned the landline companies for the submarine ventures.

Although Scudamore's plan contemplated, at least in theory, personnel passing from the telegraph companies to the Post Office, the backers of nationalization took it for granted that the top officials of the Post Office would take over the management of the new telegraph service. The only important exception was that Culley moved over to the position of chief engineer, made necessary by the lack of technical prowess in the Post Office. This enormous collective faith in the capacity of the Post Office managers to supervise the national telegraph service went hand in hand with the belief that in this way the managerial and technical know-how of the telegraph companies could continue to flow out to the submarine companies[117] and favour their expansion. In 1871 William Stratford Andrews, secretary of the United, moved over to the same position in the Indo-European Telegraph Company, running a mixed line (cables and landline) from Great Britain to India. The chairman was Grimston, who had held the same position in the Electric and from 1867 was

---

[116] Colin Hampsted, "The Early Years of Oceanic Telegraphy: Technology, Science and Politics", *IEE Proceedings*, 136/A6, 1989, pp. 297–305.

[117] Simone Müller, "The Transatlantic Telegraphs and the Class of 1866 – The Formative Years of Transnational Networks in Telegraphic Space, 1858–1884–89", *Historical Social Research*, 35/1, 2010, pp. 237–259.

on the board of the Atlantic. Weaver, who had been the last secretary of the Electric, carried out the same function in the Indo-European before the arrival of Andrews and afterwards in the Anglo-American Telegraph Company. Lastly, Edward Bright, the historic secretary of the Magnetic and then the British & Irish, joined his brother as a freelance consultant for cable companies.[118]

6.6   A new turning point

The five-year period between 1865 and 1870 was a breaking point with the past. The Post Office had been mandated to deal with the domestic telegraph service given the sorry state it was in. This meant a radical change both in the management and the business strategies adopted. The new managers, almost all from the Post Office, immediately carried out the reforms the public had been calling for, namely an extremely economical flat charge and the building of new lines where needed. The national service was at once transformed from exclusive to universal, and at the same time, was swallowed up within the multiservice structure of the Post Office. The long-term features consolidated in over 30 years were destined to change as the new constitutive choices took effect. With nationalization, Great Britain had adapted to the dominant model in Europe, thus managing to access the TU. At the same time, however, the public acquisition of the private network at a decidedly high price led to financing of the global cable network which was indirect and therefore without any political responsibility. Just when the domestic telegraphs were becoming more European, the global network was becoming more and more British.

---

[118] J. Wagstaff Blundell, *Manual of Submarine Telegraph Companies*, London, published by the author, 1872; Roberts, *Distant Writing*, pp. 302–315.

Conclusion

The British telegraph sector is a significant example of Victorian capitalism. From a technological point of view, it is collocated across the two industrial revolutions. It was introduced by entrepreneurial innovators in the first part of nineteenth century to then be developed and extended in the later phase, especially for the maritime use, by scientists with an interest in practical applications. From the point of view of business, British telegraphs incarnated all the contradictions of European liberalism in the 1860s, which, while aspiring the values of free competition did not disdain the pragmatism of cartel agreements and natural monopolies.

There emerged in the course of the evolution of the telegraph market a vicious circle which ended by holding back its full development. Set in motion by the iron will of Cooke and consolidated by Ricardo's acute business strategies, landline telegraphs were born as a private monopoly with high economic and legal barriers to discourage any rivals from entering. The few that managed to do so always based part of their business strategy on exploiting technological niches, legal or territorial, left free by the Electric. And in fact, in the mid 1850s the market seemed to be opening up to free competition, thanks to the presence of companies like the Magnetic and the Submarine, small but with great potential. But for business to expand, there had to be changes in strategy. The entry en masse of so many small shareholders took place in this moment of expansion. Its effect was to stimulate the distribution of the profits and discourage long-term infrastructural investments. The birth of the duopoly in the late 1850s followed by an oligopolistic cartel within a few years were the results of the policies attempting to maintain the status quo. What at a certain point caused the vicious circle whereby the greater the profits, thanks to the long-term strategies, the more the shareholders intent on abandoning them in order to keep dividends high?

This phenomenon can be explained over the long period by considering some specific aspects of the British telegraphs. First of all, the market was always very asymmetric, both for demand and supply. On one side, the oligopolistic regime already brought disadvantages to its users, with its high tariffs and poor distribution of correspondence and news. On the other, it must be remembered that the main users of the service were also a substantial part of the shareholders of the main companies. Top figures from the railways, business circles and the press were on the boards of all the companies.

In the second place, the Government's attitude was to say the least striking, as was that of the entire political class totally uninterested until the cartel agreement came, and then touting nationalization. Probably the absence of norms regulating competition and new entries contributed to the cartel and the strengthening of the oligopoly.

Thirdly, a paradoxically crucial role was played by the regulation forbidding overhead lines in urban areas and along main highways. Its negative effect was that it sent infrastructural costs far higher than in other industrialized countries and impeded the entry of newcomers. Its positive effect was that that it encouraged R&D investment into waterproof lagging for submarine cables, generating a know-how that was to prove essential for the submarine sector.

Lastly, landline telegraphy, a private service, was crushed between the efficiency of the Penny Post, a Government service, and the powerful expansion of submarine telegraphy in the wake of the Empire. The presence of an efficient and economical postal service certainly hampered the possibilities of the telegraph service right from the start and confined it to an exclusive usership and the speed factor. The increasing need to interact politically and economically with the far-flung corners of the Empire had led the Government to promote the laying of enormous intercontinental cables. In this way, from the late 1850s on submarine telegraphy began to absorb the human and financial capital which had previously gone into land telegraphs.

Altogether, these four elements contributed to render the telegraph market changeable and uncertain, and for this reason explain the loss of long-term investments and preference for short-term profits. These same conditions favoured the sea change – the rapid nation-alization of the land system – and also promoted a notable increase in British submarine telegraphy, which in the 1870s began setting up an authentic global network of telecommunications.

Bibliography

Primary sources

British Telecom Archives, London

Electric Telegraph Company (TGA), 1846–1855

TGA 1/1, Chairman's books, 1850–1857.

TGA 1/2, Chairman's order books, 1848–1850.

TGA 1/3/1, Board meeting book, 1846–1850.

TGA 1/3/2, Board meeting book, 1850–1853.

TGA 1/4, Secretaries' orders, 1856–1859.

TGA 1/5/1, Assignment of a licence and portion of patent right comprised in a patent as to improvements in electric telegraphs, 1846.

TGA, 1/5/2, Assignment of three Irish patents for inventions of improvements in electric telegraphs, 1846.

TGA 1/5/3, Assignment of three Scotch patents for inventions of improvements in electric telegraphs, 1846.

TGA 1/6, Shareholders' books, 1846–1856.

TGA 1/7/1, Quarterly meetings of Superintendents Reports, 1854–1857.

TGA 1/7/2, Minutes of general meetings, 1854–1865.

TGA 1/8/1, Memoranda book 1, 1852–1855.

TGA 1/8/2, Memoranda book 2, 1854–1855.

TGA 1/8/3, Memoranda book 3, 1855–1856.

TGA 1/8/4, Memoranda book 4, 1856–1856.

TGA 1/9, Articles of agreement between Electric Telegraph Company and Universal Private Telegraph Company, 1861.

TGA 1/10/1, Rules and regulations book, 1863.

TGA 1/10/2, Rules and regulations book, 1863.

TGA 1/11, Rents, stations and list of charges books, nd.

TGA 1/12, Maps of lines and stations, 1852.

TGA 1/13, Charts of the Company's telegraph system in Great Britain,1853.

TGA 1/14, Maps of the telegraph lines of Europe, 1860.

TGA 2/1, Half yearly reports, 1851–1872.

International Telegraph Company (TGC), 1853–1855

TGC 1/1, Proposals for a submarine cable between the UK and the Netherlands, 1852–1853.

TGC 1/2, Royal charter to form a company for laying submarine cable to Holland, 1852.

TGC 1/3/1, Board agenda book 1, 1853–1858.

TGC 1/4, Board meeting books, 1853–1855.

Electric & International Telegraph Company (TGE), 1855–1870

TGE 1/1/1, Board Agenda book, 1856–1859.
TGE 1/1/2, Board Agenda book, 1857–1858.
TGE 1/2, Half yearly meetings, 1861–1868.
TGE 1/3/4, Minute book 5, 1867–1869.
TGE 1/3/5, Minute book 6, 1869–1870.
TGE 1/4, Stock register books, 1856.
TGE 1/5, Telegraph Bill correspondence books, 1866–1868.
TGE 1/6, Opposition to Telegraph Bill books, 1860–1869.
TGE 1/7/1, Scale of charges book, 1863.
TGE 1/7/2, Scale of charges book, 1866.
TGE 1/8, Tables of charges books, 1862.
TGE 1/9, Porterage books, 1867–1869.
TGE 1/10, Rules for messengers books, 1865.
TGE 1/11, Letter books, 1856–1859.
TGE 1/12/1, Bradford memoranda book, 1855–1857.
TGE 1/12/2, Bradford memoranda book, 1858.
TGE 1/12/3, Bradford memoranda book, 1859–1861.
TGE 1/13, Accountant memoranda books, 1858–1863.
TGE 1/15, Inventions by C.F. Varley, 1873.
TGE 1/16, Hours of attendance tables, 1856.
TGE 1/17, Signalling rules, nd.
TGE 1/19, Instructions for printing circuit connections, 1867.
TGE 1/20/1, 1860 chart, 1860.
TGE 1/20/2, 1867 chart, 1867.
TGE 1/21, Maps of lines and stations in Great Britain, 1859.
TGE 1/22, Maps of lines and stations in Europe, 1859.
TGE 2/1, Rates and Assessments books 1–10, 1856–1870.
TGE 2/2, Accounts books, 1855–1856.
TGE 2/3, Returns traffic book, 1866–1870.
TGE 2/4, Staff books, 1853–1858.
TGE 2/5, Accounts and engineers reports, 1867.
TGE 3/1/1, Superintendent letter book, 1868–1869.
TGE 3/1/2, Letter book, 1869–1870.
TGE 3/2, Arrangements and revenues books, 1865–1870.
TGE 3/3, Supply of press telegraphy books, 1869–1905.
TGE 3/4, Special Wire agreements, 1866–1869.
TGE 3/5, Intelligence reports books, 1865–1869.

British Electric Telegraph Company (TGB), 1850–1853

TGB 1/1/1, Legal opinion, 1853–1856.

British Telegraph Company (TGL), 1853–1857

TGL 1/1/1, Deed of settlement of the British Telegraph Company, 1854–1856.

TGL 1/1/2, Supplemental deed of settlement for the dissolution of the British Telegraph Company, 1857.

English and Irish Magnetic Telegraph Company (TGN), 1852–1857

TGN 1/1/1, Memoranda, 1852.
TGN 1/1/2, Deed of settlement of the English and Irish Magnetic Telegraph Company, 1852–1854.
TGN 1/1/3, Correspondence from the company's agent, 1852.

British and Irish Magnetic Telegraph Company (TGF), 1857–1870

TGF 1/1, Abstracts of minutes, 1864–1868.
TGF 1/2, Correspondence on meteorological scheme, 1860–1862.
TGF 1/3, Wayleave agreements for Daventry, 1864.
TGF 1/4, Inventions shown at international exhibition, 1862.
TGF 2/1, Inland and overseas charges for messages, 1865.
TGF 2/3, Traffic returns, 1869.

United Kingdom Electric Telegraph Company (TGH), 1860–1870

TGH 1/1/1, Consents, 1862–1865.
TGH 1/1/2, Correspondence, 1868.
TGH 1/2/1, Circuit returns from offices in the metropolitan, English and Scottish districts, 1868.
TGH 1/2/2, Logs of messages received from and forwarded to company stations via the Norwegian cable, 1869–1870.
TGH 1/2/3, Logs of messages forwarded from and received by company stations via the Danish cable, 1869–1870.
TGH 2/1/1, Directors' reports, statements and balance sheets, 1851–1873.
TGH 2/2/1, Commercial telegram rates, charges and refundments, 1860–1868.

London District Telegraph Company (TGG), 1859–1870

TGG 1/1, Balance sheets, 1859–1868.

Universal Private Telegraph Company (TGJ), 1861–1870

TGJ 1/1, Common Seal books, 1861–1870.
TGJ 1/2/1, Agreements with railway companies, 1861–1868.
TGJ 1/2/2, Agreements with road trustees, 1862–1865.
TGJ 1/3/1, Seal register and dividend list, 1863–1868.
TGJ 1/3/2, Seal register and dividend list, 1869.
TGJ 1/4, Letter books, 1862–1871.
TGJ 1/5, Papers relating to Professor Charles Wheatstone as a company director and shareholder, 1861–1866.
TGJ 2/1/1, Ledger, 1861–1865.
TGJ 2/1/2, Ledger, 1865–1870.

TGJ 2/2/1, Cash book, 1861–1867.
TGJ 2/2/2, Cash book 1868–1870.

Post Office

POST 30/191C, Adaption of the wires of telegraph companies to the Post Office system 1869.
TCB 211/1, Extracts from newspapers, 1838–1860.
TCB 211/2, Extracts from newspapers, 1860–1862.
TCB 211/3,Extracts from newspapers, 1864–1867.
TCB 305/7, Telegraph Acts, 1863–1943.
TCB 305/8, The Telegraph Acts, 1863–1906.
POST 8/43, Reports by the Committee of Inquiry regarding private telegraph companies' accounts 1868–1869.
POST 8/44, Diary of Mr Richardson, member of the Post Office Committee of Inquiry.
POST 30/363E, Costs for telegraph undertakings of various railways and telegraph companies acquired by the Post Office.
POST 30/202C, Transfer of telegraph companies to the Post Office: Electric and International Telegraph Company part 1, 1847–1870.
POST 30/203A, Transfer of telegraph companies to the Post Office: Electric and International Telegraph Company part 2, 1868–1880.
POST 30/311B, Purchase by the Post Office of the Universal Private Telegraph Company, part 1, 1864–1870.
POST 30/312, Purchase by the Post Office of the Universal Private Telegraph Company, part 2, 1870–1877.
POST 30/281C, Acquisition of the Economic Telegraph Company by the Post Office 1864–1875.
POST 30/248B, Acquisition of the United Kingdom Electric Telegraph Company by the Post Office, 1868–1876.
POST 30/217, Purchase by the Post Office of the British & Irish Magnetic Telegraph Company, 1868–1906.
POST 30/306, Acquisition of telegraphs by the Post Office.
POST 30/201B, Inspection visit of the telegraphs transferred to the State in Ireland.
POST 30/200C, Arrangements for engineering staff from telegraph companies acquired by the Post Office.
POST 30/203B, Transfer of telegraph clerks to the Post Office from railway companies 1870–1874.
POST 30/195C, Tenders for supply of telegraphic apparatus, 1869.
POST 30/194B, Tenders for supply of materials for the construction of telegraph poles 1869.
POST 30/269D, Consideration of using Metropolitan subways and sewers to convey telegraph wires, 1868–1875.
POST 30/211B, Pole rents of the London and Provincial Telegraph Company 1859–1870.

POST 30/287C, Agreements relating to the provision of free transmission of telegrams for railway companies, 1851–1875.

TCB 280/8, Foreign and telegraph agreements, correspondence and memoranda, book 1, 1868–1904.

TCB 280/9, Foreign telegraph agreements, with copy correspondence and memoranda, book 2, 1849–1905.

TCB 280/10, Foreign telegraph agreements, book number 1, 1869–1905.

TCB 280/11, Foreign telegraph agreements, book number 2, 1851–1906.

TCB 280/12, Foreign telegraph agreements, book number 3, 1869–1906.

TCB 305/2, Acts of Parliament relating to telegraph companies, 1844–1861.

TCB 305/3, Telegraph Companies Acts, 1853–1862.

TCB 305/6, Telegraph Acts and Regulations. Mr Benton, 1863–1935.

TCB 305/11, The Electric Telegraph Company. Secretary, 1844–1863.

TCB 305/21 Telegraph Acts, 1863–1908.

TCB 305/22 Telegraph Acts, 1863–1911.

TCB 305/23, Telegraph Acts, 1863–1911.

TCB 305/24, Telegraph Acts, 1863–1911.

TCB 305/25, Telegraph Acts, 1863–1911.

TCB 305/26, Telegraph Acts, 1863–1911.

TCB 305/53, Report of Select Committee on Telegraph Bill 1869.

IET Archives, London

SC Mss 007, Sir William Fothergill Cooke Papers, 1836–1880.

Sc Mss 009, Papers of John Joseph Fahie, 1863–1918.

SC Mss 010, Papers of Messers Highton, 1840–1848.

SC Mss 015, Papers of Edward Davy, 1836–1847.

SC Mss 020, Papers of Sir Charles Bright, 1863–1870.

London Science Museum (Files related to object)

T/1867–37, Original telegraph cable to Euston Square and Camden.

T/1876–1272, Five-needle Cooke telegraph.

T/1876–1275, Two-needle telegraphs.

T/1884–114, Wheatstone's ABC telegraph receiver, 1858.

T/1884–95, Gothic case double-needle telegraph instrument, 1851.

T/1888–198, Bain's chemical telegraph, 1850.

T/1923–263, Portable ABC set.

T/1951–165, Wheatstone original printing telegraph of 1841.

T/1964–82, Wheatstone ABC telegraph set.

T/1862–74, Henley's discharging rod.

T/1876–1285, Henley's magneto electric double-needle telegraph, 1848.

T/1884–167, Wheatstone's electrical whistle telegraph.

T/1939–139, Machine used for covering wires with silk and cotton for electrical purposes, 1840.

T/1949–323, Single-needle telegraph made by Electric Telegraph Company, 1846.

T/1949–323, Double-needle telegraph, 1840.

T/2008–8817, Telegraph, Great Eastern railway, 1867.

T/1980–935, Engraving: The Magnetic Needle Telegraph.

T/1985–1137, Engraved handbill: Electro-Magnetic Telegraph on the Great Western Railway, invented by Professor Wheatstone.

T/1939–139, Machine used for covering wires with silk and cotton for electrical purposes, 1840.

T/1949–323, Single-needle telegraph, 1846.

T/1949–323, Double-needle telegraph, 1840.

T/2008–8817, Telegraph, Great Eastern Railway, from the Electric and International Telegraph Company, 1867.

T/1995–227, Papers about the life and work of W.T. Henley and his companies, 1850–1960.

T/1860–60, Gisborne and Smith's insulator for telegraph wires mounted on short pole.

T/1860–61, Six vulcanite insulators for telegraph wires mounted on short pole.

T/1894–280, Insulators used for Morse's telegraph.

T/1876–1291, Siemen's direct writing Morse inker, 1862.

T/1876–1418, Cooke and Wheatstone's ABC Telegraph Receiver, 1840.

T/1897–45, Dial telegraph and machine, also receiver.

T/1965–126, Small ABC telegraph of French manufacture, 1855.

T/1876–1289, Cooke and Wheatstone's ABC telegraph transmitter, 1840.

T/1884–102, Cooke and Wheatstone's ABC telegraph transmitter, 1840.

T/1884–103, Cooke and Wheatstone's ABC telegraph transmitter, 1840.

T/1884–106, Cooke and Wheatstone's ABC Telegraph Transmitter, 1840.

T/1884–107, Cooke and Wheatstone's ABC telegraph transmitter, 1840.

T/1884–108, Cooke and Wheatstone's ABC telegraph transmitter, 1840.

T/1884–109, Cooke and Wheatstone's ABC telegraph transmitter, 1840.

T/1884–110, Wheatstone's ABC telegraph transmitter, 1858.

Newspapers (1836–1871)

Bath Chronicle and Weekly Gazette
Belfast News-letter
Birmingham Daily Post
Blackburn Standard
Bradford Observer
Bucks Herald
Bury and Norwich Post
Caledonian Mercury
Chelmsford Chronicle
Cheltenham Chronicle
Cork Examiner
Daily Post
Devizes and Wiltshire Gazette

Dublin Evening Mail
Dundee Advertiser
Dundee, Perth and Cupar Advertiser
Durham County Advertiser
Essex Standard
Fife Herald
Freeman's Journal
Glasgow Herald
Gloucester Journal
Hampshire Advertiser
Hampshire Telegraph
Hereford Times
Huddersfield Chronicle
Illustrated Times
Kendal Mercury
Kentish Gazette
Leamington Spa Courier
Leeds Intelligencer
Leeds Mercury
Leicester Chronicle
Leicester Journal
Leicestershire Mercury
Liverpool Daily Post
Liverpool Mercury
Lloyd's Weekly Newspaper
London Daily News
London Evening Standard
London Standard
Manchester Courier and Lancashire General Advertiser
Manchester Times
Morning Chronicle
Morning Post
Newcastle and Tyne Mercury
Newcastle Courant
Newcastle Journal
Norfolk Chronicle
Northern Whig
Northampton Mercury
Nottinghamshire Guardian
Pall Mall Gazette
Preston Chronicle
Reading Mercury
Reynolds's Newspaper
Royal Cornwall Gazette
Stamford Mercury
Stirling Observer

Sussex Advertiser
Times
Waterford Mail
Westmorland Gazette
Worcestershire Chronicle
York Herald
Yorkshire Gazette
Yorkshire Post and Leeds Intelligencer

Secondary sources

Adley Charles C., *Anglo-Indian Telegraphs or Communication with London, in Six Hours*, Calcutta, Sanders, Cones & Co., 1854.
Anderson James, *Existing and Projected Telegraph Routes to India Considered*, London, Metchim & Son, 1868.
Anderson James, "Statistics of Telegraphy", *Journal of the Statistical Society of London*, 35/3, 1872, pp. 272–326.
Atlantic Telegraph Company, *A History of Preliminary Experimental Proceedings and a Descriptive Account of the Present State & Prospects of the Undertaking*, London, Jarrold and Sons, 1857.
Atlantic Telegraph: Its History, from the Commencement of the Undertaking in 1854, to the Sailing of the "Great Eastern" in 1866, London, Bacon and Co., 1866.
Baines Frederick Ebenezer, *Forty Years at the Post Office. A Personal Narrative.* London, Richard Bentley and Sons, 1895.
Bentham Jeremy, "On the Liberty of the Press, and Public Discussion, Addressed to the Spanish People", in John Bowring (ed.), *The Works of Jeremy Bentham*, Part VII, Edinburgh, William Tait, 1839.
Blavier Édouard Ernest, *Cours théorique et pratique de télégraphie électrique*, Paris, De Lacroix-Comon, 1857.
Blavier Édouard Ernest, *Nouveau traité de télégraphie électrique*, Paris, Eugene Lacroix, 1866–1867.
Blundell J. Wagstaff, *Manual of Submarine Telegraph Companies*, London, Published by the Author, 1872.
Bois Victor, *La télégraphie électrique*, Paris, Librarie L. Hachette, 1853.
Bond R., *The Handbook of the Telegraph. Being a Manual of Telegraphy, Telegraph Clerks' Remembrances and Guide to Candidates for Employment in the Telegraph Service*, London, Virtue Brother and Co., 1864.
Bond R., *Handbook of the Telegraph. Being a Manual of Telegraphy, Telegraph Clerks' Remembrances and Guide to Candidates for Employment in the Telegraph Service*, London, Lockwood & Co., 1873.
Breguet Louis François Clément, *Manuel de télégraphie électrique*, Paris, L. Hachette, 1862.
Bright Charles, *Submarine Telegraphs. Their History, Construction, and Working*, London, Crosby Lockwood and Son, 1898.
Bright Charles, *Imperial Telegraphic Communication*, London, P.S. King & Son, 1911.

Bright Edward Brailsford and Bright Charles, *The Life Story of the Late Sir Charles Tilston Bright, Civil Engineer*, Westminster, Archibald Constable and Co., n.d. (2 volumes).

Bureau International des Administrations Télégraphiques, *La legislation télégraphique*, Berne, Imprimerie Rieder & Simmen, 1876.

Carême F. and Raynaud J., *Le télégraphe automatique de sir Charles Wheatstone*, Paris, Dunod, 1876.

Chadwick Edwin, "On the Economy of Telegraphy as Part of a Public System of Postal Communication", *Journal of the Society of Arts*, 15/745 1 March 1867, pp. 221–236.

Chadwick Edwin, "On the Railway Reform in Connection with a Cheap Telegraphic Post and a Parcel Post Delivery", *Journal of the Society of Arts*, 15/778 18 October 1867, pp. 717–728.

Clark Latimer, *Birmingham Wire Gauge. A Paper Read at the British Association for the Advancement of Science at Dundee, September 1867*, London, E. & F. N. Spon, 1867.

Cooke Thomas Fothergill, *Authorship of the Practical Electric Telegraph of Great Britain*, London, W.H. Smith and Son, 1867.

Cooke William Fothergill, *Telegraphic Railways; or the Single Way, Recommended by Safety, Economy, and Efficiency, under the Safeguard and Control of the Electric Telegraph*, London, Simkin, Marshall & C.,1842.

Cooke William Fothergill, *The Electric Telegraph. Was It Invented by Professor Wheatstone? Part I, Pamphlets of 1854–56*, London, W.H. Smith and Son, 1857.

Crotch Arthur, *The Hughes and Baudot Telegraphs*, London, S. Rentell & Co. Limited, 1908.

Culley Robert S., *A Handbook of Practical Telegraphy*, London, Longman, Green, Roberts and Green, 1867.

Culley Robert S., *A Handbook of Practical Telegraphy*, London, Longman, Green and Co., 1885.

Dood George, *Railways, Steamers and Telegraphs. A Glance at Their Recent Progress and Present State*, London-Edinburgh, W. & R. Chambers, 1868.

Douglas John Christie, *A Manual of Telegraph Construction: The Mechanical Elements of Electric Telegraph Engineering*, London, Charles Griffin and Company, 1877.

Du Moncel Théodose, *Considérations Nouvelles sur l'électromagnétisme et ses applications*, Paris, J. Claye, 1853.

Du Moncel Théodose, *Traité theorique et pratique de télégraphie eléctrique, a l'usage des employes télégraphistes, des ingenieurs, des constructeurs et des inventeurs*, Paris, Gauthier-Villars, 1864.

Elliot Arthur D., *The Life of George Joachim Goschen, First Viscount Goschen, 1831–1907*, London, Longmans Green, 1911.

Fahie John Joseph, *A History of Electric Telegraphy to the Year 1837*, London, E & F.N. Spon, 1884.

Field Henry M., *The Story of the Atlantic Telegraph*, New York, Charles Scribner's Sons, 1893.

Figuier Louis, *Exposition et Histoire des Principales Découvertes Scientifiques Modernes*, Paris, Langlois et Leclerq-Victor Masson, 1851.

Fleming John Ambrose, *Fifty Years of Electricity. The Memories of an Electrical Engineer*, London, Iliffe & Sons Limited, 1921.

Fleming John Ambrose, *Memories of a Scientific Life*, London-Edinburgh, Marshall, Morgan & Scott Ltd, 1934.

Garratt G.R.M., *One Hundred Years of Submarine Cables*, London, His Majesty's Stationery Office, 1950.

Gavarret Jules, *Télégraphie électrique*, Paris, Victor Masson et fils, 1861.

Gisborne Francis, *Telegraphic Communication with India*, London, Edward Stanford, 1868.

Goldsmid Frederic John, *State and Prospects of the Existing Indo-European Telegraph: Being a Reply to Recent Public Assertions on the Subject of Telegraphic Communication with India*, London, Edward Stanford, 1868.

Goldsmid Frederic John, *Telegraph and Travel. A Narrative of the Formation and Development of Telegraphic Communication between England and India, under the Orders of Her Majesty's Government, with Incidental Notices of the Countries Traversed by the Lines*, London, Macmillan and Co., 1874.

Gooch Daniel, *Diaries of Sir Daniel Gooch, Baronet*, London, Kegan Paul, Trench Trubner & Co., 1892.

Government and the Telegraphs. Statement of the Case of the Electric & International Telegraph Company against The Government Bill for Acquiring the Telegraphs, London, Effingham Wilson, 1868.

Hand-book to Electric Telegraph, London, W. Scales, Shoreditch, Clarck Warwick Lane, 1847.

Head Francis Bond, *Stockers and Pokers. The London and North Western Railway, The Electric Telegraph and the Railway Clearing House*, London, John Murray, Albemarle Street, 1861.

Helps Arthur, *Life and Labours of Mr. Brassey*, London, Bell and Daldy, 1872.

Highton Edward, *The Electric Telegraph: Its History and Progress*, London, John Weale, 1852.

Hoskiaer Otto Valdemar, *Laying and Repairing of Electric Telegraph Cables*, London, E. & F.N. Spon, 1878.

Jevons William Stanley, *Methods of Social Reforms and Other Papers*, London, Macmillan and Co., 1883.

Lardner Dionysius, *The Electric Telegraph Popularised*, London, Walton and Maberly, 1855.

Laurencin Paul, *La télégraphie terrestre, sous marin, pneumatique*, Paris, J. Rothschild, 1877.

The London Anecdotes for All the Readers, London, David Bogue, 1848.

Matteucci Carlo, *Manuale di telegrafia elettrica*, Torino, Unione Tipografica Editrice, 1861.

Maxwell Herbert, *Life of the Right Honourable William Henry Smith M.P.*, London, William Blackwood and Sons, 1894.

Mc Calum David, *The Globotype Telegraph: A Recording Instrument by which Small Coloured Balls Are Released One by One and Made to Pass over a Series of*

Inclined Planes by the Force of Their Own Gravity, London, Longman, Brown Green and Longmans, 1856.

Mercadier Ernest Jules Pierre, *Traité élémentaire de télégraphie électrique*, Paris, G. Masson, 1880.

Meyer Hugo Richard, *The British State Telegraphs. A Study of the Problem of a Large Body of Civil Servants in a Democracy*, New York, Macmillan & Co., 1907.

Miege Bernard, *Vade-mecum pratique de télégraphie électrique a l'usage des employés du télégraphe*, Paris, J. Hetzel et C.ie, 1855.

Mill John Stuart, *Principles of Political Economy*, Vol. II, Boston, Charles C. Little & James Brown, 1848.

Mullaly John, *The Laying of the Cable or the Ocean Telegraph Being a Complete and Authentic Narrative of the Attempt to Lay the Cable across the Entrance to the Gulf of St. Lawrence and of the Three Atlantic Telegraph Expeditions of 1857 and 1858*, New York, Appleton and Company, 1858.

Newall Robert S., *Facts and Observations Relating to the Invention of the Submarine Cable and the Manufacture and Laying of the First Cable between Dover and Calais in 1851*, London, E. & F. Spon, 1882.

Ogilvie A.M., "The Rise of the English Post Office", *The Economic Journal*, 3/11, 1893, pp. 443–457.

Ogilvie A.M., "The State in Its Relations to Telegraphs and Telephones", *The Economic Journal*, 18/72, 1908, pp. 602–609.

Ogilvie A.M., "A New History of the Post Office", *The Economic Journal*, 23/89, 1913, pp. 137–141.

Parkinson J.C., *The Ocean Telegraph to India. A Narrative and a Diary*, Edinburgh and London, William Blackwood and Sons, 1870.

Phipson T., *A Manual of the Various Electro-Magnetic Telegraphs at Present in Use*, London, J. Gilbert- J. M'Combe, 1847.

Preece William Henry, *On the Railways Telegraphs and the Application of Electricity to the Signalling and Working of Trains*, London, William Clowes and Sons, 1863.

Preece William Henry and Sivewright James, *Telegraphy*, London, Longmans Green and Co., 1876.

Prescott George B., *History, Theory and Practice of the Electric Telegraph*, Boston, Ticknor and Fields, 1860.

Progress Peter, *The Rail and the Electric Telegraph Comprising a Brief History of Former Modes of Travelling and Telegraphic Communication*, London, R. Yorke Clarke and Co., 1848.

Ronalds Francis, *Descriptions of an Electrical Telegraph*, London, R. Hunter, 1823.

Sabine Robert, *The Electric Telegraph*, London, Virtue Brothers & C., 1867.

Sabine Robert, *History and Progress of the Electric Telegraph*, London Lockwood & C., 1872.

Saward George, *The Trans-Atlantic Submarine Telegraph: A Brief Narrative of the Principal Incidents in the History of the Atlantic Telegraph Company*, London, printed for private circulation, 1878.

Bibliography 219

Seeligmann T., *Indian Rubber and Gutta Percha: A Complete Practical Treatise on Indian Rubber and Gutta Percha in Their Historical, Botanical, Arboricultural, Mechanical, Chemical and Electrical Aspects*, London, Scott Greenwood, 1903.

Shaffner Tal P., *The Telegraph Manual: A Complete History and Description of the Semaphoric, Electric and Magnetic Telegraphs of Europe, Asia, Africa and America, Ancient and Modern*, New York, Pudney & Russel, 1859.

Smith Adam, *An Inquiry into the Nature and Causes of the Wealth of Nations*, Oxford, Clarendon Press, 1880.

Smith Willoughby, *A Résumé of the Earlier Days of Electric Telegraphy*, London, Hayman Brothers and Lilly, 1881.

Smith Willoughby, *The Rise and Extension of Submarine Telegraphy*, London, J.S. Virtue & Co., 1891.

Suarez Saavedra Antonio, *Tratado de telegrafía, Historia universal de la telegrafía*, Barcelona, Imprenta de Jaime Jepus, 1880–1882.

Ternant Ludovic A., *Transmission des signaux par les câbles sous-marins*, Parigi, Ducher, 1875.

Ternant Ludovic A., *Le siphon-recorder et le curb-sender automatique*, Paris, G. Masson, 1882.

Tuck Henry, *The Railway Shareholder Manual, or the Practical Guide to All the Railways in the World, Completed, in Progress, and Projected; Forming an Entire Railways Synopsis Indispensable to All Interested in Railway Locomotion*, London, Effingham Wilson, 1848.

Turnbull Laurence, *The Electro-Magnetic Telegraph*, Philadelphia, A. Hart, 1853.

Walker Charles V., *Electric Telegraph Manipulation*, London, George Knight, 1850.

West Charles, *The Story of My Life, by the Submarine Telegraph*, London, John Capman and Company, 1859.

Wheatstone Charles T., "An Account of Some Experiments to Measure the Velocity of Electricity and the Duration of the Electric Light", *Philosophical Transactions of the Royal Society*, 124, 1834, pp. 583–591.

Whishaw Francis, *The Railways of Great Britain and Ireland*, London, John Weale, 1842.

Wilson George, *Electricity and the Electric Telegraph*, London, Longman, Brown, Green and Longmans, 1852.

Wilson George, *The Progress of the Telegraph*, Cambridge, Macmillan and Co., 1859.

Contemporary literature

Ahvenainen Jorma, *The Far Eastern Telegraphs: The History of Telegraphic Communications between the Far East, Europe and America Before the First World War*, Helsinki, Finnish Academy of Science and Letters, 1981.

Ahvenainen Jorma, *The European Cable Companies in South America Before the First World War*, Helsinki, Finnish Academy of Science and Letters, 2004.

Ahvenainen Jorma, "The International Telegraph Union: The Cable Companies and the Governments", in Finn Bernard, Yang Daqing (eds), *Communications under the Seas: The Evolving Cable Network and Its Implications*, Cambridge, MA, London, MIT Press, 2009, pp. 61–79.

Allen Michelle Elizabeth, *Cleansing: Sanitary Geographies in Victorian London*, Athens, OH, Ohio University Press, 2008.

Appleyard Rollo, *Pioneers of Electrical Communications*, London, Macmillan and Co., 1930.

Badenoch Alexandre, Fickers Andreas, "Introduction: Europe Materializing? Toward a Transnational History of European Infrastructures", in Alexandre Badenoch and Andreas Fickers (eds), *Materializing Europe: Transnational Infrastructures and the Project of Europe*, Basingstoke, Hampshire and New York, Palgrave Macmillan, 2010, pp. 1–23.

Baglehole Kenneth Charles, *A Century of Service. A Brief History of Cable and Wireless LTD. 1868–1968*, London, Cable & Wireless Limited, 1969.

Bailey Michael R. (ed.), *Robert Stephenson; The Eminent Engineer*, Aldershot, Ashgate, 2003.

Baker Edward Cecil, *Sir William Preece, FRS, Victorian Engineer Extraordinary*, London Hutchinson & Co., 1976.

Balbi Gabriele, "Studying the Social History of Telecommunications. Between Anglophone and Continental Tradition", *Media History*, 15/1, 2009, pp. 85–101.

Balbi Gabriele, "Telecommunications", in Peter Simonson, Janice Peck, Robert T. Craig, John P. Jackson (eds), *Handbook of Communication History*, London and New York, Routledge, 2012, pp. 209–222.

Balbi Gabriele, *Network Neutrality. Switzerland's Role in the Genesis of the Telegraph Union, 1855–1875*, Bern, Peter Lang, 2014.

Balbi Gabriele, Calvo Spartaco, Fari Simone, Richeri Giuseppe, "'Bringing Together the Two Large Electric Currents that Divide Europe': Switzerland's Role in Promoting the Creation of a Common European Telegraph Space, 1849–1865", *ICON*, 15, 2009, pp. 61–80.

Balbi Gabriele, Fari Simone, Calvo Spartaco, Richeri Giuseppe, "Specialità svizzere. L'influenza della Confederazione elvetica sull'origini dell'Unione Telegrafica, 1855–1875", *Tst: Transportes, Servicios y telecomunicaciones*, 25, 2013, pp. 150–175.

Balbi Gabriele, Fari Simone, Calvo Spartaco, Richeri Giuseppe, "Swiss Specialties: Switzerland's Role in the Genesis of the Telegraph Union, 1855–1875", *Journal of European Integration History*, 19/2, 2013, pp. 207–225.

Balbi Gabriele, Winterhalter Cecilia (eds), *Antiche novità. Una guida transdisciplinare per interpretare il vecchio e il nuovo*, Napoli-Salerno, Orthotes Editrice, 2013.

Barbier Dominique, Bertho-Lavenir Catherine, *Histoire des medias: de Diderot à Internet*, Paris, A. Colin, 2003.

Barton Roger Neil, "New Media. The Birth of Telegraphic News in Britain 1847–68", *Media History*, 16/4, 2010, pp. 379–406.

Bayly Christopher Alan, *The Birth of the Modern World 1780–1914. Global Connections and Comparisons*, Malden, Oxford, Victoria, Blackwell Publishing, 2004.

Beauchamp Ken, *History of Telegraphy*, Stevenage, IEE, 2001.

Bektas Yasup, "The Sultan's Messenger: Cultural Constructions of Ottoman Telegraphy, 1847–1880", *Technology and Culture*, 41/3, 2000, pp. 669–696.

Bell Duncan S.A., "Dissolving Distance: Technology, Space, and Empire in British Political Thought, 1770–1900", *The Journal of Modern History*, 77, 2005, pp. 523–562.

Bell Duncan S.A., "Empire and International Relations in Victorian Political Thought", *The Historical Journal*, 49/1, 2006, pp. 281–298.

Beniger James, *The Control Revolution. Technological and Economic Origins of the Information Society*, Cambridge, MA, Harvard University Press, 1989.

Bowers Brian, *Sir Charles Wheatstone FRS, 1802–1875*, London, Science Museum, 1975.

Bowers Brian, "Early Electric Telegraphs", in Frank A.J.L. James (ed.), *Semaphore to Short Waves*, Proceedings of a conference on the technology and impact of early telecommunications held at the Royal Society for the Encouragement of Arts, Manufactures and Commerce, 29 July 1996, organised by the British Society for the History of Science and the RSA, London, RSA, 1998, pp. 20–25.

Bowers Brian, *Sir Charles Wheatstone FRS 1802–1875*, London, IEE, 2001.

Boyce Robert, "Submarine Cables as a Factor in Britain's Ascendancy", in Michael North (ed.), *Kommunikationsrevolutionen. Die neunen Medien des 16. und 19. Jahrhunderts*, Köln-Weimar-Wein, Böhlau Verlag, 1995, pp. 81–100.

Boyce Robert W.D., "Imperial Dramas and National Realities: Britain, Canada and Struggle for a Pacific Telegraph", in *Imperialism: Crisis and Panic in the Indian Empire, c. 1830–1920*, Basingstoke, Palgrave Macmillan, 2010.

Bühlmann Elisabeth, *La ligne Siemens: la construction du télégraphe indoeuropéen 1867–1870*, Frankfurt, Peter Lang, 1999.

Buliung Ron N, "Wired People in Wires Places: Stories about Machines and the Geography of Activity", *Annals of the Association of American Geographers*, 101/6, 2011, pp. 1365–1381.

Burns Russell, *Communications: An International History of the Formative Years*, London, IEE, 2004.

Choudhury Deep Kanta Lahiri, *Telegraphic Imperialism: Crisis and Panic in the Indian Empire, c. 1830–1920*, Basingstoke, Palgrave Macmillan, 2010.

Clayton Howard, *The Atmospheric Railways*, Lichfield, private publication, 1966.

Clow D.G., "Pneumatic Tube Communication Systems in London", *The Newcomen Society for the Study of the History of Engineering and Technology Transactions*, 66, 1994–1995, pp. 97–120.

Coase R.H., "Rowland and the Penny Post", *Economica*, 6/24, November 1939, pp. 423–435.

Coates Vary T., Finn Bernard, *A Retrospective Technology Assessment: Submarine Telegraphy. The Transatlantic Cables of 1866*, San Francisco, San Francisco Press, 1979.

Codding George Arthur Jr, *The International Telecommunication Union. An Experiment in International Cooperation*. Leiden, E.J. Brill, 1952.

Colley Robert, "Railways and the Mid-Victorian Income Tax", *The Journal of Transport History*, 24/1, 2003, pp. 78–102.

Colquhoun Kate, *A Thing in Disguise. The Visionary Life of Joseph Paxton*, London, Harper Perennial, 2004.

Cooksey Alfred J.A., *Electric Telegraphs in Dorset*, Dorset, Dorset County Council, 1975.

Cookson Gillian, *The Cable. The Wire That Changed the World*, Stroud, Tempus, 2003.

Cookson Gillian, "Submarine Cables: Novelty and Innovations 1850–1870", *Transactions of the Newcomen Society*, 76/2, 2006, pp. 207–220.

Cookson Gillian, Hempstead Colin, *A Victorian Scientist and Engineer. Fleming Jenkin and the Birth of Electrical Engineering*, Aldershot-Brookfield, Ashgate, 2000.

Cowhey Peter F., "The International Telecommunications Regime: The Political Roots of Regimes for High Technology", *International Organization*, 44/2, 1990, pp. 169–199.

Cox John G., *The Achievements and Failings of a Great Railway Developer*, London Railway & Canal Historical Society, 2008.

Crawley Chetwode, *From Telegraphy to Television. The Story of Electrical Communications*, London, Frederick Warne & Co., 1931.

Dalling G., "David Waddington. A Great Survivor", *The Great Eastern Railway Society Journal*, 19/20, 1978, pp. 20–21.

David Paul A., "Clio and the Economics of QWERTY", *American Economic Review*, 75/2, 1985, pp. 332–337.

David Paul A., "Path Dependence: A Foundational Concept for Historical Social Science", *Cliometrica: The Journal of Historical Economics and Econometric History*, 1/2, 2007, pp. 145–175.

de Andrade Martins Roberto, "Resistance to the Discovery of Electromagnetism: Oersted and the Symmetry of the Magnetic Field", in Bevilaqua Fabio and Giannetto Enrico (eds), *Volta and the History of Electricity*, Pavia/Milano, Universitá degli Studi di Pavia, Hoepli, 2003, pp. 245–265.

de Cogan Donard, "Dr. E.O.W. Whitehouse and the 1858 Trans-Atlantic Cable", *History of Technology*, 10, 1985, pp. 1–15.

Denman, R.P.G., *Electrical Communication, I. Line Telegraphy and Telephony, Catalogue of the Collections in the Science Museum South Kensington*, London, His Majesty's Stationery Office, 1926.

Dent N.J.H., "The Tensions in Liberalism", *The Philosophical Quarterly*, 38/153, 1988, pp. 481–485.

Dibner Bern, *The Atlantic Cable*, Norwalk, Burndy Library, 1959.

Dibner Bern, *Oersted and the Discovery of Electromagnetism*, New York, Blaisdell, 1962.

Dordick Herbert S., "The Origins of Universal Service. History as a Determinant of Telecommunications Policy", *Telecommunications Policy*, 14, 1990, pp. 223–231.

du Boff Richard B., "The Telegraph in Nineteenth-Century America: Technology and Monopoly", *Comparative Studies in Society and History*, 26/4, 1984, pp. 571–586.

Dunsheath Percy, *A History of Electrical Engineering*, London, Faber and Faber, 1962.

Durand Barthez Patrice, *Union Internationale des Télécommunications, Thèse pour le doctorat en droit, Université de Paris I – Pantheon – Sorbonne Sciences Economiques – Sciences Humaines – Sciences Juridiques*, 1979.

Durham John, *Telegraphs in Victorian London*, Cambridge, The Golden Head Press, 1959.

Dzur Albert W., "Nationalism, Liberalism, and Democracy", *Political Research Quarterly*, 55/1, 2002, pp. 191–211.

Edsall Nicholas C., *Richard Cobden Independent Radical*, Cambridge, MA, Harvard University Press, 1986.

Fari Simone, "Technology on the Wire. Technological Changes in the First Thirty Years of the Italian Telegraph Experience: Achievements and Difficulties", in Giuntini Andrea (ed.), *Communication and Its Lines. Telegraphy in the 19th Century among Economy, Politics and Technology*, Prato, Istituto di Studi Storici Postali, 2004, pp. 135–158.

Fari Simone, *Una Penisola in Comunicazione. Il servizio telegrafico dall'Unità alla Grande Guerra*, Bari, Cacucci Editore, 2008.

Fari Simone, *The Formative Years of the Telegraph Union*, Cambridge, Cambridge Scholar Publishing, 2015.

Fari Simone, "Financing Telegraph Infrastructure: The Case of Great Britain (1850–1900)", in Youssef Cassis, Giuseppe De Luca, Massimo Florio (eds), *Infrastructure Finance in Europe: Insights into the History of Water, Transport and Telecommunication*, Oxford, Oxford University Press, forthcoming.

Fari Simone, Balbi Gabriele, Richeri, Giuseppe, "Telecommunications Italian Style: The Shaping of the Constitutive Choices (1850–1914)", in Anna Guagnini, Luca Molá (eds), *Italian Technology from the Renaissance to the Twentieth Century*, "History of Technology" (book-series) n. 32, London-New Delhi-New York-Delhi, Bloomsbury, 2014, pp. 235–258.

Federer J. Peter, "Advances in Communication Technology and Growth of the American Over-the-Counter Markets, 1876–1929", *The Journal of Economic History*, 68/2, 2008, pp. 501–534.

Field Alexander James, "The Magnetic Telegraph, Price and Quantity Data, and the New Management of Capital", *The Journal of Economic History*, 52/2, 1992, pp. 401–413.

Finer Samuel Edward, *The Life and Times of Sir Edwin Chadwick*, London, Methuen, 1952.

Finn Bernard S., Yang Daqing (eds), *Communication under the Seas: The Evolving Cable Network and Its Implications*, Cambridge, MA, MIT Press, 2009.

Flichy Patrice, *Une histoire de la communication moderne: espace public et vie privée*. Paris, La Decouverte, 1991.

Foreman-Peck James, "Competition, Cooperation and Nationalization in the Nineteenth Century Telegraph System", *Business History*, 31/3, 1989, pp. 81–102.

Franksen Ole Immanuel, *H. C. Oersted – A Man of the Two Cultures*, Birlerod, Strandsberg Forlag, 1981.

Freebody J.W., *Telegraphy*, London, Sir Isaac Pitman & Sons Ltd, 1958.

Friedrich Alexander, "Metaphorical Anastomoses: The Concept of 'Network' and Its Origins in the Nineteenth Century", in Neumann Birgit, Ansgar Nünning (eds), *Travelling Concepts for the Study of Culture*, Berlin/Boston, Walter de Gruyter, 2012, pp. 119–144.

Garbade Kenneth D., Silber William L., "Technology, Communication and the Performance of Financial Markets; 1840–1975", *The Journal of Finance*, 33/3, 1977, pp. 819–832.

Gassler Carrie, "Some Problems in the Development of the Communications Industry", *The American Economic Review*, 35/4, 1945, pp. 585–606.

Gillings Annabel, *Brunel. Life & Times*, London, Haus Publishing, 2006.

Gitelman Lisa, *Always Already New: Media, History and the Data of Culture*, Cambridge, MA, MIT Press, 2006.

Gitelman Lisa, Pingree Geoffrey B. (eds), *New Media, 1740–1915*, Cambridge, MA, MIT Press, 2003.

Giuntini Andrea, *Le meraviglie del mondo: il sistema internazionale delle comunicazioni nell'Ottocento*, Prato, Istituto di studi storici postali, 2011.

Golden Catherine J., *Posting It. The Victorian Revolution in Letter Writing*, Gainesville, University Press of Florida, 2009.

Gordon Scott, "The London Economist and the High Tide of Laissez Faire", *Journal of Political Economy*, 63/6, 1955, pp. 461–488.

Greig J. Michael, "The End of Geography? Globalization, Communications, and Culture in the International System", *Journal of Conflict Resolution*, 46/2, 2002, pp. 225–243.

Griset Pascal, *Entreprise, technologie et souveraineté : les télécommunications transatlantiques de la France*, Paris, Editions Rive Droite, 1996.

Hadfield Charles, *Atmospheric Railways*, Gloucester, Alan Sutton Publishing Limited, 1985.

Hamill Lynne, "The Social Shaping of British Communications Networks Prior to the First World War", *Historical Social Research*, 35/1, 2010, pp. 260–286.

Hampf Michaela, Müller-Pohl Simone (eds), *Global Communication Electric. Business, News and Politics in the World of Telegraphy*, Frankfurt and New York, Campus, 2013.

Hannah Leslie, "Logistic, Market Size, and Giant Plants in the Early Twentieth Century: A Global View", *The Journal of Economic History*, 68/1, 2008, pp. 46–79.

Harper John, *Monopoly and Competition in British Telecommunications. The Past, the Present and the Future*, London, Pinter, 1997.

Harris Christina Phelps, "The Persian Gulf Submarine Telegraph of 1864", *The Geographical Journal*, 135/2, 1969, pp. 169–190.

Harris David, "European Liberalism in the Nineteenth Century", *The American Historical Review*, 60/3, 1955, pp. 501–526.

Hazlewood Arthur, "The Origin of the State Telephone Service in Britain", *Oxford Economic Papers*, New Series, 5/1, 1953, pp. 13–25.

Headrick Daniel R., *The Tools of Empire: Technology and European Imperialism in the Nineteenth Century*, New York, Oxford University Press, 1981.

Headrick Daniel R., *The Tentacles of Progress: Technology Transfer in the Age of Imperialism, 1850–1940*, New York and Oxford, Oxford University Press, 1988.

Headrick Daniel R., *The Invisible Weapon. Telecommunications and International Politics*, New York-London, Oxford University Press, 1991.

Headrick Daniel R., *When Information Came of Age. Technologies of Knowledge in the Age of Reason and Revolution, 1700–1850*, London & New York, Oxford University Press, 2000.

Headrick Daniel R., "A Double-Edged Sword: Communications and Imperial Control in British India", *Historical Social Research*, 35/1, 2010, pp. 51–65.

Headrick Daniel R., Griset Pascal, "Submarine Telegraph Cables: Business and Politics, 1838–1939", *Business History Review* 75/3, 2001, pp. 543–578.

Hempstead Colin A., "The Early Years of Oceanic Telegraphy: Technology, Science and Politics", *IEE Proceedings*, 136/A6, 1989, pp. 297–305.

Hempstead Colin A., "Representations of Transatlantic Telegraphy", *Engineering Science and Education Journal*, December 1995, pp. S17–S25.

Henry Katherine, *Liberalism and the Culture of Security*, Tuscaloosa, University of Alabama Press, 2011.

Henry Nancy, Schmitt Cannon, *Victorian Investments. New Perspectives on Finance and Culture*, Bloomington, Indiana University Press, 2009.

Hilder Sam, *From Smoke Signals to Telstar*, Hemel Hampstead, Herts, Model Aeronautical Press, 1966.

Hills Jil, *The Struggle for Control of Global Communication: The Formative Century*, Urbana, University of Illinois Press, 2002.

Hoag Christopher, "The Atlantic Cable and Capital Market Information Flows", *The Journal of Economic History*, 66/2, 2006, pp. 342–353.

Hobsbawm Eric, *Age of Capitalism 1848–1875*, London, Hachette, 2010.

Hochfelder David, "A Comparison of the Postal Telegraph Movement in Great Britain and the United States, 1866–1900", *Enterprise & Society*, 1, 2000, pp. 739–761.

Hubbard Geoffrey, *Cooke and Wheatstone and the Invention of the Electric Telegraph*, London, Routledge & Kegan, 1965.

Hugill Peter J., *Global Communications since 1844. Geopolitics and Technology*, Baltimore, Johns Hopkins University Press, 1999.

Hunt Bruce J., "The Ohm Is Where the Art Is: British Telegraph Engineers and the Development of Electrical Standards", *Osiris*, 2nd Series, 9, 1994, pp. 48–63.

Huurdeman Anton A., *The Worldwide History of Telecommunications*, London, J. Wiley, 2003.

John Richard R., "Private Enterprise, Public Good? Communications Deregulation as a National Political Issue, 1839–1851", in Andrea Giuntini (ed.), *Communication and Its Lines. Telegraphy in the 19th Century among*

Economy, Politics and Technology, Prato, Istituto di Studi Storici Postali, 2004, pp. 33–56.

John Richard R., "Telecommunications", *Enterprise and Society*, 9, 2008, pp. 507–520.

John Richard R., *Network Nation. Inventing American Telecommunications*, London and Cambridge, The Belknap Press of Harvard University Press, 2010.

Johnson Paul, *Making the Market. Victorian Origins of Corporate Capitalism*, Cambridge, Cambridge University Press, 2010.

Kahan, Alan S. *Liberalism in Nineteenth-Century Europe: The Political Culture of Limited Suffrage*, Basingstoke and New York, Palgrave Macmillan, 2003.

Kaukianen Yrjö, "Shrinking the World: Improvements in the Speed of Information Transmission, c. 1820–1870", *European Review of Economic History*, 5/1, 2001, pp. 1–28.

Kennedy Paul M., "Imperial Cable Communications and Strategy, 1870–1914", in *The English Historical Review*, 86/341, 1971, pp. 728–752.

Kent David, "Containing Disorder in the Age of Equipoise: Troops, Trains and the Telegraph", *Social History*, 38/3, 2013, pp. 308–327.

Kieve Jeffrey, *The Electric Telegraph, A Social and Economic History*, Devon, David and Charles Newton Abbot, 1973.

Kusserow Karl, "Technology and Ideology in Daniel Huntington's *Atlantic Cable Projectors*", *American Art*, 24/1, 2010, pp. 94–113.

Laborie Leonard, "En chair et en normes. Les participants aux conférences de l'Union internationale des télécommunications, de sa fondation à sa refondation (1865–1947)", *Flux. Cahiers scientifiques internationaux Réseaux et Territoires*, 74, 2008, pp. 92–98.

Laborie Leonard, *L'Europe mise en réseaux. La France et la coopération internationale dans les postes et les télécommunications, années 1850–années 1950*, Bruxelles, PIE Peter Lang, 2010.

Leigh Williams Frances, *Matthew Fontaine Maury. Scientist of the Sea*, New Jersey, Rutgers University Press, 1963.

Lequeux James, *François Arago, un savant généreux*, Paris, EDP-Sciences, 2008.

Levinson Paul, *Soft Edge. A Natural History & Future of the Information Revolution*, London, Routledge, 1997.

Lightman Bernard, *Victorian Popularizers of Science. Designing Nature for New Audiences*, Chicago, University of Chicago Press, 2007.

Lindley Lester G., "Watered and Control of Telegraph Rates: Early Proposals for Regulating a Public Utility", *The Journal of Economic History*, 32/1, 1972, pp. 403–408.

Lubrano AnnTeresa, *The Telegraph: How Technology Innovation Caused Social Change*, New York, Garland, 1997.

Malachuk Daniel S., *Perfection, the State and Victorian Liberalism*, New York and Basingstoke, Hampshire, Palgrave Macmillan, 2005.

Mc Closkey H.T., "Mill's Liberalism", *The Philosophical Quarterly*, 13/51, 1963, pp. 143–156.

McMahon Peter, "Communications Technology and Structural Change in the International Political Economy – The Cases of Telegraphy and Radio", *Prometheus: Critical Studies in Innovation*, 20/4, 2002, pp. 379–390.

Michie R.C., "The London Stock Exchange and the British Securities Market, 1850–1914", *The Economic History Review*, New Series, 38/1, 1985, pp. 61–82.

Mokyr Joel, "Technological Inertia in Economic History", *The Journal of Economic History*, 52/2, 1992, pp. 325–338.

Morus Iwan Rhys, "Currents from the Underworld: Electricity and the Technology of Display in Early Victorian England", *Isis*, 84/1, 1993, pp. 50–69.

Morus Iwan Rhys, "The Electric Ariel: Telegraph and Commercial Culture in Early Victorian England", *Victorian Studies*, 39/3, 1996, pp. 339–378.

Morus Iwan Rhys, "The Nervous System of Britain: Space, Time and the Electric Telegraph in the Victorian Age", *The British Journal for the History of Science*, 33/4, 2000, pp. 455–475.

Morus Iwan Rhys, "Worlds of Wonder: Sensation and the Victorian Scientific Performance", *Isis*, 101/4, 2010, pp. 806–816.

Müller-Pohl Simone, "The Transatlantic Telegraphs and the Class of 1866 – The Formative Years of Transnational Networks in Telegraphic Space, 1858–1884–89", *Historical Social Research*, 35/1, 2010, pp. 237–259.

Müller-Pohl Simone, "By Atlantic Telegraph. A Study on Weltcommunication in the 19th Century", *m&z*, 4, 2010, pp. 40–54.

Müller-Pohl Simone, "Working the Nation State: Submarine Cable Actors, Cable Transnationalism and the Governance of the Global Media System, 1858–1914", in Isabella Löhr and Roland Wenzlhuemer (eds), *The Nation State and Beyond. Governing Globalization Processes in the Nineteenth and Twentieth Century*, Berlin, Springer Verlag Berlin, 2013, pp. 101–123.

Murray Evelyn, *The Post Office*, London, New York, G.P. Putnam's Sons Ltd, 1927.

Nicholls Arthur R., *The London & Portsmouth Direct Atmospheric Railway: "A mere puff of wind"*, Stroud, Fonthill Media, 2013.

Nickles David Paul, *Under the Wire. How the Telegraph Changed Diplomacy*, Cambridge MA, Harvard University Press, 2003.

Nonnenmacher Tomas, "Law, Emerging Technology, and Market Structure: The Development of the Telegraph Industry, 1838–1868", *The Journal of Economic History*, 57/2, 1997, pp. 488–490.

Nonnenmacher Tomas, "State Promotion and Regulation of the Telegraph Industry, 1845–1860", *The Journal of Economic History*, 61/1, 2001, pp. 19–36.

Nonnemacher Tomas, "Network Quality in the Early Telegraph Industry", *Research in Economic History*, 23, 2005, pp. 61–82.

Nye David E., "Shaping Communication Networks: Telegraph Telephone Computer", *Social Research*, 64/3, 1997, pp. 1067–1091.

Olivé Roig Sebastián, *Historia de la telegrafía óptica en España*, Madrid, Ministerio de Trasporte, Turismo y Comunicaciones, 1990.

Olivé Roig Sebastián, *El nacimiento de la telecomunicación en España*, Madrid, Cuadernos de Historia de telecomunicaciones n° 4, Escuela Técnica Superior de Ingenieros de Telecomunicación, 2004.

Otter Chris, *Victorian Eye. A Political History of Light and Vision in Britain, 1800–1910*, Chicago, University of Chicago Press, 2008.

Overman Michael, *Understanding Telecommunications*, Portsmouth, Eyre & Spottiswide, 1974.

Park David W., Janowsky Nicholas W., Jones Steve (eds), *The Long History of New Media: Technology, Historiography and Newness in Context*, New York, Peter Lang, 2011.

Peacock Alf, "George Leeman and York Politics 1833–1880", in C.H. Feinstein (ed.), *York, 1831–1981: 150 Years of Scientific Endeavour and Social Change*, York, William Sessions Limited, 1981, pp. 234–254.

Perry Charles Richard, "Frank Ives Scudamore and the Post Office Telegraphs", *Albion: A Quarterly Journal Concerned with British Studies*, 12/4, 1980, pp. 350–367.

Perry Charles Richard, *The Victorian Post Office. The Growth of a Bureaucracy*, Wooldbridge, The Royal Historical Society-The Boydell Press, 1992.

Perry Charles Richard, "The Rise and Fall of Government Telegraphy in Britain", *Business and Economic History*, 26/2, 1997, pp. 416–425.

Pinch Trevor, Bijker Wiebe E., "The Social Construction of Facts and Artefacts: Or How the Sociology of Technology Might Benefit Each Other", *Social Studies of Science*, 14/3, 1984, pp. 419–424.

Pitt Douglas C., *The Telecommunications' Function in the British Post Office. A Case Study of Bureaucratic Adaptation*, Westmead, Farnborough, Saxon House, 1980.

Plotz John, *Portable Property. Victorian Culture on the Move*, Princeton, Princeton University Press, 2008.

Pomper Philip, "The History and Theory of Empires", *History and Theory*, 44/4, 2005, pp. 1–27.

Potter Simon J., "Webs, Networks, and Systems: Globalization and the Mass Media in the Nineteenth and Twentieth Century British Empire", *Journal of British Studies*, 46/3, 2007, pp. 621–646.

Read Donald, *The Power of News. The History of Reuters*, New York, Oxford University Press, 1999.

Richeri Giuseppe, "The Media Amidst the Enterprises, the Public and the State", *Studies in Communication Sciences*, 6/2, 2006, pp. 131–143.

Roach John, "Liberalism and the Victorian Intelligentsia", *Cambridge Historical Journal*, 13/1, 1957, pp. 58–81.

Roberts Steven, *Distant Writing. A History of the Telegraph Companies in Britain between 1838 and 1868*, London, Steven Roberts, 2006–2012, p. 6. Unpublished in hardcopy, available at http: //distantwriting.co.uk/, accessed 8 August 2013.

Rosenberg Nathan, *Exploring the Black Box: Technology, Economics, and History*, Cambridge UK and New York, Cambridge University Press, 1994.

Ross David, *George and Robert Stephenson: A Passion for Success*, Stroud, History Press, 2010.

Scott J.D., *Siemens Brothers 1858–1958. An Essay in the History of Industry*, London, Weindenfeld and Nicolson, 1958.

Shahvar Soli, "Iron Poles, Wooden Poles: The Electric Telegraph and the Ottoman: Iranian Boundary Conflict, 1863–65", *British Journal of Middle Eastern Studies*, 34/1, 2007, pp. 23–42.

Shapiro Salwyn J., "John Stuart Mill, Pioneer of Democratic Liberalism in England", *Journal of the History of Ideas*, 2/4, 1943, pp. 127–160.

Shelangoskie Susan, "Anthony Trollope and the Social Discourse of Telegraphy After Nationalization", *The Journal of Victorian Culture*, 14/1, 2009, pp. 72–92.

Shiers George (ed.), *The Electric Telegraph. An Historical Anthology*, New York, Arno Press, 1977.

Showalter Dennis, "Soldiers into Postmasters? The Electric Telegraph as an Instrument of Command in the Prussian Army", *Military Affairs*, 37/2, 1973, pp. 48–52.

Silverstone Roger, "What's New about New Media?", *New Media and Society*, 1, 1999, pp. 10–12.

Slater Ernest, *One Hundred Years. The Story of Henley's*, London, Henley's Telegraph Works Company Limited, 1937.

Smith Ken, *Stephenson Power: The Story of George and Robert Stephenson*, Newcastle upon Tyne, Tyne Bridge Publishing, 2003.

Solymar Laszlo, *Getting the Message. A History of Telecommunications*, New York, Oxford University Press, 1999.

Souden David, *Voices over the Horizon. Tales from Cable & Wireless*, London, Cable and Wireless, 1999.

Spinner Thomas J., *George Joachim Goschen: The Transformation of a Victorian Liberal*, Cambridge, Cambridge University Press, 1973.

Sraffa Piero, Maurice Dobb (eds), *The Works and Correspondence of David Ricardo, Biographical Miscellany*, Vol. 10, Indianapolis, Liberty Fund, 2005.

Stacey Tom, *Thomas Brassey: The Greatest Railway Builder in the World*, London, Stacey International, 2005.

Staff Frank, *The Penny Post: 1680–1918*, London, Lutterworth Press, 1964.

Standage Tom, *The Victorian Internet. The Remarkable Story of the Telegraph and the Nineteenth Century's On-line Pioneers*, London, Weidenfeld & Nicolson, 1998.

Starr Paul, *The Creation of the Media. Political Origins of Modern Communications*, New York, Basic Books, 2004.

Steel Gordon John, *A Thread across the Ocean. The Heroic Story of the Transatlantic Cable*, London, Simon & Schuster, 2002.

Sterling Christopher, Bernt Phyllis, Weiss Martin B.H., *Shaping American Telecommunications: A History of Technology. Policy and Economics*, Mahwah, New York, Lawrence Erlbaum Associates, 2006.

Stoddart Coggeshall Ivan, *An Annotated History of Submarine Cables and Overseas Radiotelegraphs: 1851–1934*, in Donard de Cogan (ed.), Norwich, School of Information Systems (UEA), 1993.

Storey Graham, *Reuter's Century 1851–1951*, London, Max Parish, 1951.

The Telcon Story 1850–1950, London, The Telegraph Construction & Maintenance Co. Ltd, 1950.

Thomson Janet, *The Scot Who Lit the World. The Story of William Murdoch, Inventor of Gas Lighting*, Glasgow, self-published, 2003.

Tomory Leslie, *Progressive Enlightenment. The Origins of the Gaslight Industry, 1780–1820*, Cambridge, MA, MIT Press, 2012.

Trainer Matthew, "The Role of Patents in Establishing Global Telecommunications", *World Patent Information*, 29, 2007, pp. 352–362.

Van der Vleuten Erik, "Towards a Transnational History of Technology: Meanings, Promises, Pitfalls", *Technology and Culture*, 49/4, 2008, pp. 974–994.

Vaughan Adrian, *Brunel – An Engineering Biography*, Surrey, Ian Allan Publishing, 2006.

Wallsten Scott, "Returning to Victorian Competition, Ownership, and Regulation: An Empirical Study of European Telecommunications at the Turn of the Twentieth Century", *The Journal of Economic History*, 65/3, 2005, pp. 693–722.

Watson Garth, *The Civils*, London, Thomas Telford, 1988.

Wenzlhuemer Roland, "Editorial – Telecommunication and Globalization in the Nineteenth Century", *Historical Social Research/Historische Sozialforschung*, 35/1, 2010, pp. 7–18.

Wenzlhuemer Roland, "Globalization, Communication and the Concept of Space in Global History", *Historical Social Research/Historische Sozialforschung*, 35/1, 2010, pp. 19–47.

Wenzlhuemer Roland, "The History of Standardization in Europe", *European History Online* (EGO) (2010), available at http://www.ieg-ego-eu/wenzl-huemer-2010-en, accessed 4 May 2014.

Wenzlhuemer Roland, *Connecting the Nineteenth-Century World. The Telegraph and Globalization*, Cambridge UK, Cambridge University Press, 2013.

Williams M.B., *History and Evolution of Telecommunications*, typewritten, BT Library, 1974.

Wilson Charles, *First with the News: The History of W.H. Smith, 1792–1972*, London, Jonathan Cape, 1985.

Wilson Geoffrey, *The Old Telegraphs*, London, Phillimore, 1976.

Winseck Dwayne R., "Globalizing Telecommunications and Media History: Beyond Methodological Nationalism and the Struggle for Control Model of Communication History", in Hampf Michaela, Müller-Pohl Simone (eds), *Global Communication Electric. Business, News and Politics in the World of Telegraphy*, Frankfurt/New York, Campus Verlag, 2013, pp. 35–62.

Winsek Dwayne R., Pike Robert M., *Communication and Empire: Media, Markets, and Globalization, 1860–1930*, Durham, Duke University Press, 2007.

Winsek Dwayne R., Pike Robert M., "Communication and Empire: Media Markets, Power and Globalization, 1860–1910", *Global Media and Communication*, 4/1, 2008, pp. 7–37.

Yates JoAnne, "The Telegraph's Effect on Nineteenth Century Markets and Firms", *Business and Economic History*, 15, 1986, pp. 149–163.

Index of Names